# The Book On Attention

## Focused Realms and Navigating Attention in a Distracted World

The Book On Series

# J. M. Raines

Published by The Book On Publishing, 2025.

First edition. October 25, 2025

Website: https://thebookon.ca

Substack: https://thebookonpublishing.substack.com/

While every precaution has been taken in the preparation of this book, the publisher assumes no responsibility for errors or omissions or damage resulting from the use of the information contained herein.

The Book On Attention: Focused Realms and Navigating Attention in a Distracted World

First edition. October 25, 2025

ISBN: 978-1-997909-42-2

Written by J. M. Raines

## Other Books in The Book On Series

# Table of Contents

# Chapter 1: Attention in the Age of Distraction

By picking up this book, you chose to focus on these words instead of many other options. In that singular act lies a capacity that defines human consciousness more fundamentally than language, memory, or reason. Attention is the invisible architecture of experience itself, the selection mechanism that determines what enters your awareness and what remains in the shadows. Yet in an era characterized by relentless sensory bombardment, this most precious cognitive resource has become paradoxically both more valuable and more vulnerable than at any previous moment in human history. We stand at an inflection point where the mechanisms designed to capture our attention have grown so sophisticated, so pervasive, and so deeply integrated into the fabric of daily existence that most people navigate their days in a state of chronic distraction, their attention fragmented across dozens of simultaneous streams, never fully present to anything at all. This crisis of attention extends far beyond individual productivity or digital wellness concerns; it represents nothing less than an existential challenge to human agency, autonomy, and the capacity for meaning-making in the modern world.

Consider the mathematics of the attention economy that now governs much of contemporary life. Every notification that chimes on your device, every autoplay video that begins without permission, every carefully engineered scroll mechanism on social platforms, these are not accidents of design, but calculated interventions crafted through billions of dollars in research and development. Silicon Valley employs some of the world's most brilliant minds in persuasive technology, behavioral psychology, and neuroscience, all focused on a single objective: maximizing engagement metrics by capturing and retaining user attention for as long as possible. The economic model is elegantly simple and profoundly consequential. Attention translates directly into advertising revenue, which means that human consciousness itself has become the raw material being mined, refined, and sold in the digital marketplace. A 2019 analysis by Microsoft revealed that the average person now receives approximately 63.5 notifications daily from various applications, each one representing a micro-interruption that fragments continuous thought. The cumulative effect resembles death by a thousand cuts, where sustained focus becomes impossible not because individuals lack willpower but because the environment itself has been optimized against it. What makes this particularly insidious is the asymmetry of the conflict: individuals armed only with conscious resistance confront systems designed by teams of specialists with real-time data on what works and what does not.

The evolutionary context illuminates why we find ourselves so vulnerable to these designed distraction systems. Human attention evolved in environments radically different from those we inhabit today. For hundreds of thousands of years, our ancestors navigated savannas where survival depended on noticing subtle environmental changes, the rustle of grass that might signal a predator. This distant movement could indicate prey, the social cues from group members that determined hierarchy and cooperation. Natural selection favored brains wired to detect novelty, orient rapidly to unexpected stimuli, and maintain vigilance across multiple domains simultaneously. These attentional mechanisms served our ancestors extraordinarily well in their ecological niche. The problem emerges when these ancient circuits encounter modern technological environments specifically engineered to exploit their vulnerabilities. The notification alert, the red badge icon, and the infinite scroll each leverage deep-seated neurological tendencies toward novelty-seeking and uncertainty. Our brains did not evolve to resist the kinds of supernormal stimuli that digital platforms deliver with machine precision. The mismatch between our inherited attentional architecture and our constructed attentional environment creates a fundamental challenge that willpower alone cannot overcome.

## The Hidden Costs of Fractured Focus

What do we lose when our attention splinters across multiple streams simultaneously? The answer extends far beyond missed deadlines or incomplete tasks into the realm of human experience itself. Attention functions as the gateway to consciousness, determining what achieves the status of reality in our subjective experience. Psychologist William James famously wrote in 1890 that "my experience is what I agree to attend to," capturing an essential truth about the relationship between attention and the construction of our inner lives. When attention becomes chronically fragmented, the quality of experience itself degrades in subtle and profound ways. Neuroscientific research reveals that multitasking, more accurately described as rapid task-switching, imposes significant cognitive costs. A seminal 2009 study from Stanford University found that heavy media multitaskers performed worse than light multitaskers on tests of attention control, even when they believed themselves to be more efficient. The researchers discovered that chronic multitaskers had difficulty filtering out irrelevant information, struggled to manage their working memory effectively, and showed impaired ability to switch tasks compared to those who focused on single activities. Perhaps most troubling, they appeared less aware of their own cognitive deficits, trapped in an illusory superiority regarding their multitasking abilities.

The degradation extends into the emotional and relational dimensions of human life with consequences that resist easy quantification but emerge unmistakably in lived experience. When you sit across from someone whose eyes repeatedly dart to their phone, when you find yourself unable to watch a film without checking messages, when you realize you cannot recall the drive home because your mind wandered throughout the journey, these moments reveal attention's erosion. Clinical psychologist Adam Gazzaley and cognitive neuroscientist Larry Rosen document what they term "distraction sickness," characterized by anxiety, reduced productivity, and decreased life satisfaction directly attributable to chronic attention fragmentation. The inability to sustain focus creates a feedback loop: diminished attention leads to reduced accomplishment, which generates anxiety and stress, which further impairs attentional control. Meanwhile, a genuine human connection requires a specific form of sustained, undivided attention that becomes increasingly rare. Martin Buber's concept of the "I-Thou" relationship, an authentic encounter with another person as a whole being rather than an object to be used, depends absolutely on presence, on the capacity to give another person the gift of your full attention. When that capacity atrophies, relationships flatten into transactional exchanges, and the depth of human connection becomes impossible to achieve or sustain.

Yet the crisis encompasses dimensions beyond individual cognition and relationships, extending into the collective capacity for democratic discourse and shared reality construction. A functioning democracy requires citizens capable of sustained attention to complex problems, nuanced thinking about difficult trade-offs, and deliberation that moves beyond tribal reflexes. These cognitive capacities depend fundamentally on the ability to focus, to read deeply rather than skim, to think slowly and carefully about essential questions rather than react instantly to provocative stimuli. Political scientist Yochai Benkler and his colleagues documented how attention manipulation has become central to information warfare, with state and non-state actors deliberately flooding the information ecosystem with inflammatory content designed to capture attention, provoke emotional reactions, and prevent the kind of thoughtful engagement necessary for informed citizenship. The attention economy's incentive structures amplify the most divisive, emotionally provocative content because it generates the strongest engagement signals. Algorithmic curation systems, trained to maximize watch time and click-through rates, systematically prioritize material that triggers strong reactions over content that fosters understanding. The result is a public sphere characterized by permanent outrage, tribal signaling, and the inability to sustain focus on any single issue long enough to develop a sophisticated understanding or build coalitions for meaningful action.

## The Paradox of Abundance and Poverty

We find ourselves in a strange position: living amid information abundance while experiencing attention poverty. Never before in human history has so much knowledge been so readily accessible, yet this accessibility creates its own challenges. The paradox lies in the fact that information only becomes knowledge through the application of sustained attention, careful reading, patient reflection, and integration with existing understanding that transforms raw data into genuine insight. When attention fragments, information remains merely potential, a vast ocean through which we skim without ever diving deep enough to retrieve anything of substance. Cal Newport, computer science professor and productivity theorist, distinguishes between shallow work, tasks that can be performed while distracted, and deep work, cognitively demanding activities that require sustained focus and produce valuable output. His research indicates that the capacity for deep work has become both rarer and more useful in the modern economy. Knowledge workers who develop the ability to focus intensely create disproportionate value precisely because such focus has become uncommon. Meanwhile, constant access to information creates an illusion of knowledge without its substance, a phenomenon psychologists term the "fluency effect," in which easy access to information misleads people into believing they possess understanding they do not actually have.

The temporal dimension of attention reveals another critical aspect of the contemporary crisis. Human consciousness naturally oscillates across different time horizons, from immediate sensory experience to long-range planning and reflection on the past. Attention can focus on the present moment, anticipate future possibilities, or explore memory and meaning. Healthy psychological functioning requires fluidity across these temporal modes, the capacity to be fully present when presence matters, while also engaging in future-oriented planning and retrospective sense-making. The attention economy's relentless presentism, its focus on the immediate, the urgent, the now, distorts this temporal balance. Push notifications demand instant response, news cycles refresh continuously, and the fear of missing out creates pressure to remain perpetually updated. Philosopher Roman Krznaric describes how this temporal distortion creates what he calls the "tyranny of the immediate," where short-term thinking dominates and long-term concerns recede from view. Climate change, infrastructure decay, and education system reform are challenges that require sustained attention across years or decades, yet the attentional environment militates powerfully against such temporal extension. Similarly, the contemplative modes that generate meaning and wisdom, reflection on experience, consideration of values, and philosophical questioning require stepping outside the stream of immediate stimuli. When attention becomes imprisoned in the perpetual present, these deeper dimensions of human consciousness atrophy from disuse.

The economic and social stratification of attention represents perhaps the most troubling dimension of the crisis. Just as economic capital accumulates among those who already possess it, attentional capacity increasingly divides along class lines. Affluent knowledge workers hire assistants, deploy technological filters, attend expensive retreats focused on mindfulness and focus, and structure their environments to protect their attention. Meanwhile, those in precarious economic positions often work multiple jobs with irregular schedules, face constant workplace monitoring, and lack the resources or autonomy to create protected spaces for sustained focus. Psychologist Roy Baumeister's research on ego depletion, the theory that willpower operates as a finite resource that becomes exhausted through use, suggests that those facing economic stress and decision fatigue may have less capacity to resist attentional capture precisely because they confront more demands on their self-regulatory resources. The divide extends into educational contexts, where schools serving wealthy communities increasingly limit screen time and emphasize embodied learning, while schools in poorer areas often rely heavily on digital instruction. Silicon Valley executives famously send their children to screen-free Waldorf schools while building products designed to maximize engagement among the general population. This bifurcation threatens to create an attentional aristocracy possessing the cognitive capacities needed for complex work and autonomous agency. At the same time, a larger population becomes increasingly vulnerable to manipulation and incapable of the sustained focus necessary for advancement.

## Reclaiming Human Agency

Despite the formidable challenges, the capacity for attention remains fundamentally a human faculty subject to cultivation and protection. The first step toward reclamation involves recognition, understanding with clarity the nature and magnitude of the forces arrayed against sustained focus. This chapter has traced the evolutionary vulnerabilities that make human attention susceptible to technological exploitation, documented the psychological and social costs of chronic distraction, and exposed the economic incentives driving the attention economy. Awareness itself does not solve the problem, but it transforms the contest from an invisible manipulation into an explicit struggle where conscious choice becomes possible. Philosopher Harry Frankfurt's concept of "second-order desires", the capacity to evaluate and potentially change one's immediate wants, provides a framework for thinking about attentional autonomy. The question is not merely what captures your attention in any given moment, but whether you endorse that capture, whether it aligns with your deeper values and goals. Most distractions operate at the level of first-order impulses, triggered automatically by environmental cues. Second-order reflection creates space for genuine agency, the opportunity to ask whether this is how you want to spend the finite resource of your attention.

The subsequent chapters of this book will explore in detail the practical, psychological, and philosophical dimensions of attention cultivation in the modern world. We will examine the neuroscience of focus and how understanding brain function can inform better practices for sustained attention. We will investigate the design principles behind persuasive technology and develop literacy about the specific mechanisms used to capture and retain attention. We will explore contemplative traditions that offer time-tested methods for training attentional capacity, from meditation practices that strengthen meta-awareness to ritual structures that create protected temporal zones. We will consider how to redesign personal environments and social contexts to support rather than undermine sustained focus. We will grapple with questions of meaning and value, examining how attention expresses what we care about and how its allocation ultimately shapes the form of a human life. Throughout this exploration, we will resist both technological determinism, the notion that we are helpless before digital forces, and naive voluntarism, the belief that willpower alone suffices to resist sophisticated systems of attentional capture.

The stakes could scarcely be higher. Attention is not merely a psychological capacity or productivity tool; it is the fundamental mechanism through which we engage with reality, construct meaning, form relationships, and exercise agency in the world. To lose control of attention is to lose control of the very substance of life itself. Yet the challenge we face is not to retreat into pre-technological nostalgia or to reject the genuine benefits that digital connectivity provides. Instead, it is to develop sufficient awareness, skill, and structural support to use powerful technologies without being used by them, to benefit from information abundance without drowning in its torrents, to remain connected without becoming perpetually distracted. The path forward requires both individual cultivation and collective action, personal discipline and social reform, practical technique and philosophical reorientation. It demands that we take seriously what has always been true but becomes urgently visible in the current moment: that how we attend shapes who we become, and that in an age of manufactured distraction, the capacity for sustained focus represents nothing less than the preservation of human consciousness itself. This book is an invitation to that preservation, a guide to reclaiming the cognitive capacity that makes genuine thought, authentic relationships, and meaningful existence possible. The journey begins with this foundational recognition of both the challenge and the possibility, the crisis and the opportunity that define attention in the age of distraction.

The transformation required operates at multiple scales simultaneously, from the intimate sphere of personal habit to the systemic level of regulatory policy and corporate accountability. Consider how previous public health crises, such as tobacco addiction, industrial pollution, and food contamination, eventually prompted both individual behavior change and structural intervention. The attention crisis follows a similar trajectory but with greater complexity because the harms manifest psychologically rather than physically, making them harder to measure and easier to dismiss. When a cigarette causes cancer, the causal chain eventually becomes undeniable through medical evidence. When a notification interrupts deep thought, the damage disperses across thousands of micro-interruptions whose cumulative impact resists simple quantification. This invisibility has allowed the attention economy to expand largely unchecked, its externalities absorbed silently by individuals who blame themselves for lacking discipline rather than questioning the systems designed to undermine their focus.

Moreover, the challenge intensifies because digital technologies genuinely provide immense value alongside their costs. Email enables communication across distances that would have seemed magical to previous generations. Search engines place humanity's accumulated knowledge at our fingertips. Social platforms allow maintenance of relationships that geography would otherwise dissolve. The problem is not technology itself but the economic model that monetizes attention through advertising, creating incentives to maximize engagement regardless of its impact on human flourishing. Alternative models exist, subscription services, public utility frameworks, cooperative ownership structures, that would align platform incentives with user wellbeing rather than attention extraction. The question becomes whether societies can muster the political will to demand such alternatives, recognizing that attention, like clean air or safe food, is a public good that requires protection from exploitation. This recognition marks the transition from viewing distraction as a personal failing to understanding it as a structural challenge requiring a collective response. This shift opens possibilities currently foreclosed by individualistic frameworks of responsibility and blame.

# Chapter 2: The Psychology of Focus: Understanding the Mind's Eye

The human capacity for focus operates through mechanisms far more intricate than the metaphor of a mental spotlight suggests. While popular understanding frames attention as a beam of light illuminating objects in consciousness, cognitive science reveals a dynamic, multifaceted system involving distinct neural networks, competing priorities, and sophisticated filtering processes. The term "mind's eye" itself proves instructive; vision provides the dominant metaphor for attention precisely because both involve selection from vast perceptual fields. Yet unlike physical vision, which remains relatively constant in its mechanisms, attention demonstrates remarkable plasticity, adapting its operations based on context, experience, and internal states. Understanding these psychological foundations requires examining how attention emerges from neural architecture, how different attentional systems cooperate and compete, and how cognitive resources get allocated across competing demands for awareness.

At the neurological level, attention involves three functionally distinct networks distributed across the brain. The alerting network, primarily mediated by norepinephrine projections from the locus coeruleus, maintains a state of readiness to respond to incoming stimuli. This system governs arousal and vigilance, determining the baseline sensitivity with which the brain approaches its environment. The orienting network, involving the superior parietal cortex and frontal eye fields, directs attention toward specific locations or features in space. This system enables both voluntary shifts, consciously deciding to look at something, and involuntary captures, as when peripheral motion automatically draws the gaze. Finally, the executive control network, anchored in the anterior cingulate cortex and lateral prefrontal regions, resolves conflicts among competing responses and maintains focus despite distractions. Neuroscientist Michael Posner's work mapping these networks through carefully designed reaction-time experiments revealed that each system operates with distinct neurochemical substrates and can be selectively impaired or enhanced through various interventions. The critical insight emerges that focus does not represent a unitary phenomenon but rather the coordinated activity of multiple specialized systems, each contributing different aspects to the overall capacity for directed attention.

## The Architecture of Selective Processing

The brain confronts an overwhelming information problem at every moment. Conservative estimates suggest that the human sensory systems receive approximately eleven million bits of information per second from the environment, visual input from roughly one hundred million photoreceptors, auditory information from thirty thousand hair cells in each cochlea, tactile data from millions of mechanoreceptors distributed across the skin, plus olfactory, gustatory, proprioceptive, and interoceptive signals. Yet consciousness can process only about forty to fifty bits per second with focused attention. This staggering compression ratio, approximately 1:200,000, means that the vast majority of available information never reaches awareness. The brain must therefore employ sophisticated selection strategies, filtering mechanisms that determine what crosses the threshold into conscious experience and what remains forever unknown to the subjective self.

This selective processing operates through both bottom-up and top-down mechanisms, working in constant dialogue. Bottom-up attention occurs when stimulus properties themselves capture awareness, such as the sudden loud noise, the bright flash, or the unexpected touch. These features trigger automatic orienting responses hardwired through evolution to detect potential threats or opportunities. Cognitive psychologist Anne Treisman demonstrated, through ingenious experiments, that certain stimulus features, such as color, orientation, and motion, can "pop out" from visual arrays, detected in parallel across the entire visual field without requiring serial search. These pre-attentive processes operate automatically and rapidly, screening the sensory input for biologically relevant signals that warrant further processing. Top-down attention, conversely, involves voluntary control guided by goals, expectations, and prior knowledge. When searching for car keys, visual attention becomes selectively tuned to key-shaped objects, enhancing processing of relevant features while suppressing irrelevant details. Neuroimaging studies reveal that these top-down signals originate in prefrontal and parietal regions, modulating sensory cortices to bias processing toward goal-relevant information.

The interaction between these systems produces phenomena that reveal the constructed nature of conscious experience. Change blindness experiments demonstrate that dramatic alterations in visual scenes can go entirely unnoticed during brief interruptions or when attention is elsewhere. In classic studies by psychologists Daniel Simons and Christopher Chabris, observers watching a video of people passing basketballs often fail to notice a person in a gorilla suit walking directly through the scene, a phenomenon termed inattentional blindness. These findings challenge naive realism, the intuitive belief that conscious perception faithfully represents the external world. Instead, they reveal that awareness contains only what attention has selected for detailed processing. The subjective sense of a richly detailed perceptual experience across the entire visual field is an illusion, a kind of cognitive sleight of hand in which the brain creates the impression of complete representation from sparse sampling. Focus thus functions not merely as an enhancement of selected information but as the fundamental gatekeeper determining what achieves the status of reality in subjective experience.

## Cognitive Load and the Limits of Processing Capacity

Understanding focus requires confronting the stubborn fact of limited processing capacity. Cognitive psychologist George Miller's famous observation that working memory can hold roughly seven items, subsequently refined to approximately four "chunks" of information, points to fundamental constraints on how much mental content can receive simultaneous focused attention. This limitation does not reflect mere inefficiency but appears hardwired into the architecture of neural information processing. Working memory operates through sustained neural firing patterns in the prefrontal cortex that maintain information in an active, immediately accessible state. Maintaining these activity patterns requires metabolic resources, glucose and oxygen consumption, that impose thermodynamic limits on the number of representations that can remain active simultaneously. Neuroscientist Earl Miller's laboratory demonstrated that, even for highly practiced tasks, accurate parallel processing of multiple streams of information remains impossible; what appears to be multitasking actually reflects rapid switching between serial processing streams.

Cognitive load theory, developed by educational psychologist John Sweller, distinguishes between intrinsic load, the inherent complexity of the material being processed, and extraneous load, the load imposed by how information is presented or by the presence of irrelevant stimuli. When total cognitive load exceeds available working memory capacity, learning and performance degrade precipitously. This explains why attempting to master complex concepts while simultaneously managing distractions proves so ineffective: the extraneous load from irrelevant stimuli consumes working memory resources needed to process the intrinsic complexity of the target material. The practical implications extend beyond educational contexts into every domain requiring focus. Interface designers manipulate cognitive load through information architecture choices, organizing content to minimize unnecessary processing demands, allowing users to direct attention toward substantive tasks rather than navigation overhead. Writers manage the reader's cognitive load through sentence structure, paragraph organization, and semantic coherence, allowing comprehension to flow smoothly rather than imposing constant backtracking and reinterpretation demands.

The concept of attentional blink provides striking evidence of temporal processing limits. When observers monitor rapid serial visual presentation streams for target items, detecting a first target impairs the ability to detect a second target appearing within approximately two hundred to five hundred milliseconds. During this "blink" period, awareness of new items seems temporarily unavailable, even as standard visual processing continues. Psychologist Mary Potter's research revealed that this phenomenon reflects not perceptual failure but a bottleneck in consolidating representations into working memory. The attentional system requires time to stabilize a representation before it can process subsequent items for conscious report. This temporal constraint becomes particularly relevant in information-saturated environments where stimuli arrive rapidly. Social media feeds, news tickers, and notification streams all exploit temporal presentation rates that prevent adequate consolidation, creating subjective experiences of information flow without actual comprehension or retention.

## The Role of Mental Models and Expertise

Focus does not operate in isolation from knowledge; instead, domain expertise fundamentally transforms attentional capabilities. Expert chess players perceive board configurations differently than novices, not through enhanced visual acuity but through recognition of meaningful patterns accumulated through thousands of hours of practice. Psychologist Adriaan de Groot's pioneering studies demonstrated that chess masters could reconstruct entire game positions after only brief exposure, a feat impossible for beginners. Crucially, this advantage disappeared when pieces were arranged randomly rather than in legal game positions. The experts' superiority stemmed not from better general memory but from chunking meaningful configurations into higher-order representations. Similar patterns emerge across domains: radiologists detecting tumors in imaging scans, musicians sight-reading complex scores, and athletes anticipating opponents' movements. In each case, extensive experience creates mental models, organized knowledge structures that allow efficient processing of domain-relevant information.

These mental models transform attention by changing what counts as signal versus noise. For the expert, subtle features invisible to novices become salient and diagnostic. A physician listening to heart sounds attends to tonal qualities, timing relationships, and acoustic variations that reveal pathology, hearing meaningful information where an untrained listener detects only rhythmic thumping. This selective tuning represents years of pattern learning encoded in the connections between perception and interpretation. Cognitive scientist Gary Klein's naturalistic decision-making research revealed that expert performance in time-pressured, uncertain environments rarely involves deliberative comparison of options. Instead, recognition-primed decisions occur when experts rapidly identify situations as instances of familiar patterns and retrieve associated responses. This process relies critically on attention trained to detect the right features, those that distinguish one situation category from another.

The development of expertise thus represents progressive refinement of attentional focus itself. Deliberate practice, as characterized by psychologist Anders Ericsson, involves targeted work on specific skills with immediate feedback, gradually automating component processes and freeing attentional capacity for higher-level integration. A beginning pianist must consciously attend to finger placement, timing, and note reading simultaneously, a cognitive load that overwhelms working memory and produces halting, error-prone performance. Through practice, these components become automated, executable with minimal conscious monitoring, allowing attention to shift toward musical interpretation, emotional expression, and interaction with other musicians. This hierarchical automation, moving from effortful conscious control toward automatic execution, represents how attentional capacity paradoxically expands through its own focused application. The expert possesses not a greater absolute capacity but rather a more efficient processing that accomplishes more with the same limited resources.

## Motivational and Emotional Influences on Attention

Focus cannot be understood as purely cognitive machinery operating independently of motivation and emotion. What captures and sustains attention depends fundamentally on what matters to the organism, what the brain has learned to value through evolution, development, and individual experience. Neuroscientist Kent Berridge's research, which distinguishes "wanting" from "liking," reveals that dopaminergic systems governing motivation can become dissociated from hedonic systems governing pleasure. This distinction explains why attention becomes captured by stimuli that have acquired incentive salience, motivational significance, even when they no longer generate positive subjective experience. The compulsive checking of devices and the inability to disengage from social media, despite deriving little actual satisfaction, reflect motivational systems commandeering attention through learned associations between cues and rewards.

Emotional states profoundly modulate attentional priorities through multiple pathways. Anxiety narrows attention toward threat-relevant information, creating hypervigilance for potential dangers while reducing capacity for neutral or positive stimuli, a pattern adaptive when threats are genuine but maladaptive when anxiety becomes chronic and threat perception distorted. Depression, conversely, biases attention toward negative information and away from positive features, maintaining negative mood states through selective focus. Psychologist Andrew Mathews' extensive work on attentional bias in clinical anxiety disorders demonstrates that anxious individuals show facilitated detection of threat words in lexical decision tasks and difficulty disengaging from threat-relevant images. These biases operate largely outside conscious awareness, revealing how emotional states sculpt what enters consciousness without deliberate intent.

Positive emotional states influence attention differently, broadening rather than narrowing focus. Psychologist Barbara Fredrickson's broaden-and-build theory proposes that positive emotions expand the scope of attention, enabling more flexible, creative, and integrative thinking. Experimental evidence supports this: participants induced into positive mood states show enhanced performance on tasks requiring holistic processing, unusual word associations, and remote concept integration. The practical implications prove substantial; the attentional state conducive to detecting specific threats differs fundamentally from that supporting creative problem-solving or relationship building. Managing focus, therefore, requires managing emotional states, recognizing that optimal attention for different tasks may require different affective contexts. The stressed, anxious state may enhance vigilance but impairs the open, exploratory attention necessary for insight and innovation.

## The Phenomenology of Flow and Optimal Focus

Psychologist Mihaly Csikszentmihalyi's concept of flow describes a psychological state characterized by complete absorption in an activity, where action and awareness merge, self-consciousness diminishes, and time perception distorts. This state represents perhaps the pinnacle of focused attention, a condition where the usual division between observer and observed collapses into seamless engagement. Flow emerges when skill level matches challenge level in the zone of optimal difficulty: too easy produces boredom and mind-wandering, too difficult generates anxiety and fragmented attention. The carpenter shaping wood, the surgeon performing a complex procedure, the rock climber navigating a challenging route, each may enter flow when their developed capabilities meet meaningful challenges requiring complete present-moment engagement.

The neuroscience of flow reveals distinctive patterns of brain activity. Rather than increased activation across attention networks, flow often involves transient hypofrontality, reduced activity in brain regions associated with self-referential processing and explicit monitoring. Neuroscientist Arne Dietrich proposes that this deactivation of the prefrontal cortex allows more automated, implicit processing systems to operate without interference from conscious control. This counterintuitive finding suggests that optimal focus sometimes requires less, rather than more, cognitive effort, a paradox captured in the Zen archery principle of "effortless effort." The conscious mind trying too hard to control attention may actually interfere with the fluid, responsive engagement characteristic of flow. This insight challenges contemporary productivity culture's emphasis on willpower and conscious control, suggesting that sustainable high performance may depend more on creating conditions that allow natural absorption than on forcing attention through sheer determination.

The cultivation of flow states has profound implications for understanding focus as a lived psychological experience rather than merely a cognitive process. Flow provides intrinsic rewards; the activity becomes autotelic, worth doing for its own sake rather than for external outcomes. This stands in stark contrast to forced attention maintained through extrinsic motivation or fear of consequences, which feels effortful, draining, and unsustainable. The difference parallels philosopher Harry Frankfurt's distinction between being moved to act versus moving oneself to act, between attention captured by external forces and attention freely given to self-chosen engagement. The quality of focus itself differs fundamentally across these conditions, suggesting that understanding attention requires examining not only what brain regions activate but also the subjective character of attentional states and the conditions that give rise to effortless versus effortful engagement.

## Attention Training and Neuroplasticity

The brain's remarkable capacity for experience-dependent reorganization means that attentional capabilities are not fixed but trainable through systematic practice. Contemplative traditions discovered this principle millennia ago through meditation practices explicitly designed to cultivate attentional control. Contemporary neuroscience validates and extends these insights through studies that reveal structural and functional brain changes following attention-training interventions. Neuroscientist Richard Davidson's laboratory demonstrated that even brief mindfulness training produces measurable improvements in sustained attention, reduced mind-wandering, and enhanced emotional regulation. Longer-term practitioners show more dramatic changes: increased gray matter density in regions associated with attention control, enhanced functional connectivity between attention networks, and more efficient neural processing during attention-demanding tasks.

Different training approaches target distinct components of the attentional system. Focused attention meditation, maintaining awareness on a single object like the breath, strengthens sustained attention and reduces attentional lapses. Open monitoring meditation, maintaining broad awareness without fixation on particular contents, enhances the capacity to notice when attention has wandered and to redirect focus flexibly. Loving-kindness practices, directing positive intentions toward self and others, appear to shift attentional biases away from negative stimuli and enhance prosocial emotions that broaden attention. Video game training targeting specific attentional demands, like tracking multiple moving objects or rapidly responding to peripheral targets, produces transfer effects to untrained attention tasks. These findings reveal that attention is not a monolithic capacity but a collection of trainable skills that can be selectively developed through appropriately designed practice.

The mechanisms underlying these training effects involve both functional changes, more efficient processing in existing networks, and structural modifications, including increased myelination of connecting fibers and dendritic elaboration supporting enhanced synaptic connectivity. Critically, training effects depend on the principle of specificity: improvements tend to be greatest for attentional demands similar to those encountered in training. This suggests that developing robust attentional capabilities requires varied practice across multiple contexts rather than narrow training on a single task. The practical implications extend into educational design, workplace training programs, and therapeutic interventions for attention disorders. Rather than accepting attentional capabilities as fixed traits, evidence increasingly supports viewing focus as a developable skill responsive to systematic cultivation.

The psychology of focus ultimately reveals attention as a dynamic, multifaceted capacity shaped by neural architecture, cognitive constraints, emotional states, expertise development, and experiential training. Understanding how attention operates at these multiple levels, from neurotransmitter systems to experiential states, provides the foundation for developing more effective strategies to cultivate focus in an increasingly demanding attentional environment. The mind's eye proves to be not a passive receptor but an active selector, constructor, and interpreter of experience, shaped by both evolutionary heritage and individual development. Recognizing attention as trainable rather than fixed, as influenced by motivation and emotion rather than purely cognitive, and as operating through multiple specialized systems rather than a single unified spotlight transforms how we approach the challenge of maintaining focus. The subsequent chapters will build upon these psychological foundations to explore practical applications, examining how understanding attention's mechanisms enables more effective navigation of the modern attention economy and cultivation of the focused awareness necessary for meaningful accomplishment and genuine presence in our lives.

# Chapter 3: Designing for Attention: The Role of Aesthetics

The physical spaces we inhabit and the visual environments we encounter exert profound influence over our attentional states, yet these effects often remain invisible precisely because they operate so effectively. Consider the architectural choices that shape a library reading room: the diffused natural light filtering through tall windows, the balance of ceiling height to floor space, the acoustic damping that reduces ambient noise to a whisper, the color palette of muted earth tones, and the ergonomic furniture that supports sustained postural comfort. None of these elements announces itself, yet collectively they create conditions that invite sustained focus. Contrast this with the typical modern office: fluorescent lighting that cycles 60 times per second, creating subliminal flicker; open floor plans that expose workers to constant peripheral movement; bright accent walls chosen for "energy" rather than cognitive calm; chairs designed for short meetings rather than extended work sessions. The difference lies not in individual preferences or work ethics but in designed affordances, environmental characteristics that either facilitate or obstruct attentional depth. Understanding how aesthetic choices shape cognitive capacity requires examining the hidden grammar of visual design, spatial arrangement, sensory modulation, and temporal patterning that together constitute the attentional architecture of our surroundings.

The relationship between aesthetics and attention extends beyond mere pleasantness into the realm of cognitive processing efficiency. Environmental psychologist Roger Ulrich documented in hospital recovery studies that patients with views of natural landscapes required fewer pain medications, experienced fewer post-surgical complications, and were discharged on average 1 day earlier than those facing brick walls. The mechanism involves more than psychological comfort: natural visual environments contain fractal patterns, self-similar structures that appear at multiple scales, that human visual systems process with exceptional efficiency. Researcher Richard Taylor's analysis of Jackson Pollock's paintings revealed fractal dimensions matching those found in natural formations, potentially explaining their sustained visual appeal. When visual processing operates smoothly without encountering jarring discontinuities or cognitively demanding irregularities, attentional resources remain available for primary tasks rather than being consumed by environmental navigation. The aesthetic dimension thus functions not as decorative luxury but as cognitive infrastructure, either taxing or supporting the limited processing capacity available for focused work.

## The Neuroscience of Visual Harmony

Applications of color theory in designed environments demonstrate how aesthetic choices directly modulate neural arousal systems. Chromatic wavelength influences the suprachiasmatic nucleus, the brain's primary circadian pacemaker, through specialized photoreceptive ganglion cells containing melanopsin. Blue-enriched light suppresses melatonin production, increasing alertness but potentially disrupting sleep patterns when encountered in the evening. Conversely, amber wavelengths preserve melatonin secretion, supporting circadian alignment. Interior designer Iris Arnstein's work implementing circadian lighting systems in educational settings revealed improved test performance and reduced behavioral disruptions when classroom illumination shifted throughout the day, mimicking natural changes in the solar spectrum. The implications extend beyond lighting to encompass all chromatic choices: warm hues activate parasympathetic responses associated with relaxation, while cool tones enhance sympathetic arousal. However, context determines appropriateness; the color scheme facilitating creative brainstorming differs from that supporting meticulous detail work. Pharmaceutical giant Genentech discovered through workspace experiments that research laboratories benefited from cooler, higher-contrast environments that promote sustained vigilance. At the same time, collaboration spaces performed better with warmer palettes that reduce social inhibition.

Spatial proportion influences attention through perceptual processing demands that either align with or violate evolved aesthetic preferences. Mathematician and architectural theorist Nikos Salingaros argues that traditional architectural elements, columns, arches, cornices, and varied surface textures, create visual hierarchies at multiple scales, allowing the eye to rest at different focal distances while providing continual novelty without chaos. Modernist aesthetics emphasizing blank surfaces, sharp edges, and minimal ornamentation reduce this scalar complexity, potentially increasing cognitive load as visual systems search unsuccessfully for anticipated pattern variation. Neuroscientist Bevil Conway's brain imaging studies demonstrated that viewing spaces conforming to golden-ratio proportions, approximately 1.618:1, produced greater activity in reward-related brain regions than other ratios. This preference transcends cultural conditioning, appearing in populations without exposure to Western architectural traditions, suggesting biological foundations in the optimization of perceptual processing. The practical application involves understanding that aesthetic choices operate not through conscious deliberation but through preconscious perceptual systems that either facilitate or impede smooth information processing.

## Material Texture and Haptic Engagement

The sensory dimension of the materials surrounding us shapes attentional quality through channels that extend beyond vision into tactile, acoustic, and even olfactory domains. Psychologist Moshe Bar's research on visual processing demonstrated that rounded forms activate approach-related neural circuitry, while angular shapes trigger avoidance responses and heightened vigilance. This manifests in material choices: the smooth curve of wooden furniture invites relaxed engagement. At the same time, sharp-edged metal and glass maintain low-level alertness that accumulates cognitive costs over extended exposure. Japanese aesthetic philosophy recognizes this through wabi-sabi principles that value natural materials that reveal their substance, wood grain, stone texture, fabric weave, rather than synthetic surfaces that mask their composition. These natural materials create what artist and educator John Ruskin termed "vital beauty," engaging perceptual systems through organic variation that rewards sustained observation without overwhelming with excessive stimulation.

Acoustic properties of materials profoundly influence cognitive performance through mechanisms distinct from conscious noise awareness. Psychologist Nick Perham's studies on irrelevant sound effects revealed that speech, even in unknown languages, disrupts verbal working memory more severely than equally loud mechanical noise. The mechanism involves obligatory processing of speech-like patterns by language systems, creating interference with phonological rehearsal loops essential for maintaining information in working memory. Material choices determine acoustic environments: hard surfaces reflect sound, creating reverberation, while fibrous materials absorb it. The difference between a conference room with exposed concrete and glass versus one incorporating acoustic ceiling tiles, fabric wall panels, and carpeting determines whether conversation remains comprehensible or degrades into effortful processing. Architect Alexander Reichardt documented how Oslo's opera house achieved acoustic clarity, allowing whispered conversations while maintaining appropriate reverberation for orchestral performance through precisely calculated surface treatments and spatial volumes. These technical achievements in the service of aesthetic experience create conditions in which attention naturally flows toward content rather than struggling against environmental interference.

Olfactory design represents an underutilized dimension of aesthetic attention management. Neuroscientist Rachel Herz's research revealed that ambient scent influences cognitive performance and emotional states through direct pathways to limbic structures, bypassing thalamic relay stations that mediate other senses. Peppermint aroma enhances alertness and typing accuracy, while lavender promotes relaxation and may be beneficial for anxiety reduction, but it also impairs performance on vigilance tasks. Retail environments exploit these effects strategically: bakeries pump bread scent toward entrances to elicit approach behavior, while real estate agents bake cookies during open houses to create unconscious associations with home comfort. However, applications extend beyond commercial manipulation into therapeutic and performance domains. Japanese forest bathing practices, shinrin-yoku, demonstrate measurable stress reduction through exposure to phytoncides released by coniferous trees, suggesting that introducing particular plant species into designed environments might support attentional restoration through olfactory channels operating primarily outside awareness.

## Temporal Aesthetics and Rhythm

The temporal dimension of aesthetic experience shapes attention through pacing, rhythm, and strategic variation that either sustain engagement or precipitate fatigue. Film editor Walter Murch articulated how cutting rhythm influences viewer attention: rapid cuts every two to three seconds maintain perceptual novelty but prevent deeper engagement. In comparison, extended shots lasting fifteen to thirty seconds allow contemplative absorption, enabling emotional resonance. Contemporary digital media largely abandoned the latter approach, optimizing for immediate engagement rather than sustained attention, a choice reflecting economic imperatives rather than aesthetic principles. Cinematographer Roger Deakins demonstrates alternative possibilities: his work on "1917" employed extended tracking shots, creating immersive temporal flow that held audience attention through spatial continuity rather than editorial fragmentation. The aesthetic principle involves matching temporal pacing to the attentional depth appropriate to the content: information delivery benefits from varied pacing, which maintains alertness, while experiential content rewards sustained focus enabled by temporal continuity.

Architectural theorist Christopher Alexander's concept of "sequence" describes how movement through space creates temporal aesthetic experiences that shape attentional states. Japanese garden design employs this principle through carefully orchestrated revelation: paths wind to conceal and then reveal features progressively, maintaining perceptual novelty while avoiding overwhelming complexity. Each viewing station presents a composed scene, but movement between stations provides transition periods allowing assimilation before encountering subsequent compositions. Western equivalents appear in Gothic cathedral designs where the longitudinal axis draws attention forward toward distant altars, while transept crossings create moments of spatial expansion that pause this visual momentum. Museum exhibition design implements similar principles: galleries alternate between focused contemplation spaces for individual masterworks and circulation areas providing perceptual rest. These temporal rhythms align with ultradian cycles, ninety to one hundred twenty-minute periods of optimal alertness followed by rest phases, that characterize human cognitive performance.

Environmental change cycles influence attention by aligning with or conflicting with circadian rhythms that govern arousal, hormone secretion, and cognitive capacity. Chronobiologist Till Roenneberg's research revealed that circadian preferences, morningness versus eveningness, have strong genetic components that affect optimal performance timing by several hours. Yet designed environments typically ignore this variation, imposing uniform lighting and activity schedules regardless of individual chronotype. Progressive workplace designs incorporate adjustable environments that acknowledge these differences: tunable lighting that allows individual control, varied workspace configurations that support both collaborative and solitary activities, and policies that permit schedule flexibility. The aesthetic dimension involves recognizing that temporal patterns constitute environmental features as tangible as physical materials, the rhythm of lighting changes, the cadence of activity and rest periods, and the tempo of informational updates all shape attentional capacity as profoundly as spatial layout or chromatic choices.

## Digital Interface Aesthetics

Screen-based environments where contemporary attention increasingly resides demonstrate how aesthetic choices determine cognitive accessibility. Typography exemplifies this: typeface selection influences reading speed, comprehension, and perceived difficulty. Psychologist Kevin Larson's research comparing Consolas with Courier New revealed that well-designed screen fonts reduced reading time while increasing accuracy on cognitive tasks performed after reading. The mechanism involves a reduction in perceptual processing demands: well-designed letterforms exhibit high x-height, optimal character spacing, and subtle variations that distinguish similar characters, features that reduce cognitive load during character recognition. Yet economic pressures often prioritize visual novelty over functional optimization: websites employ fashionable display fonts, sacrificing readability for brand differentiation, while applications implement thin, low-contrast text that conforms to minimalist aesthetic trends, despite measurably impairing legibility for users with typical or declining vision.

Whitespace utilization determines information hierarchy and cognitive processing flow. Designer Jan Tschichold's typography principles emphasized that unmarked space functions as actively as marked elements, creating breathing room, allowing eye movement, planning, and perceptual grouping. Contemporary web design often violates these principles, filling every pixel with interactive elements, advertisements, or competing content streams. The result resembles the cluttered Victorian interiors that modernism rebelled against, visual environments overwhelming perceptual systems with excessive stimulation. Practical digital aesthetics apply subtractive discipline: Basecamp's Jason Fried advocates "attention-first design," removing features that fail to serve core purposes regardless of technical feasibility. This approach recognizes that aesthetic restraint constitutes ethical responsibility in the design of environments that others inhabit cognitively.

Motion and animation in digital interfaces create temporal aesthetic experiences that either clarify or obscure information relationships. Animation researchers Barbara Tversky and Julie Morrison established that effective motion serves specific cognitive functions: establishing object continuity across transformations, revealing causal relationships through temporal sequencing, and directing attention through deliberate movement. However, gratuitous animation, motion serving only decorative purposes, impairs comprehension by triggering involuntary attentional capture toward irrelevant stimuli. Operating system interfaces demonstrate this distinction: iOS employs animation sparingly to clarify spatial metaphors (applications "zooming" from icons, panels "sliding" from screen edges). At the same time, many Android implementations included physics-based animations whose complexity exceeded cognitive utility. The aesthetic principle involves purposeful motion aligned with cognitive support rather than visual entertainment, recognizing that every animated element creates attentional demands that must justify its processing costs.

## The Ethics of Aesthetic Design

Recognizing that aesthetic choices shape cognitive capacity raises ethical questions about designer responsibility. Urban planner Jan Gehl documented how post-war modernist city planning created hostile pedestrian environments: streets designed exclusively for vehicular traffic, blank building facades lacking human-scale visual interest, plazas too large and exposed to invite occupation. These aesthetic choices reflected ideological commitments to functionalist efficiency and monumental grandeur while ignoring human perceptual and social needs. The consequences manifested in abandoned public spaces, reduced street life, and communities fragmented by infrastructure hostile to casual encounter. Contemporary "new urbanist" movements attempt to correct these failures through design languages recovered from pre-automobile traditions: varied building facades, frequent street crossings, mixed-use zoning, and human-scale proportions. The aesthetic dimensions serve not arbitrary nostalgia but cognitive and social affordances supporting community coherence.

Corporate environments demonstrate how aesthetic manipulation can serve either humane or exploitative purposes. Casino design represents an extreme case: no windows to block circadian cues, maze-like layouts that frustrate attempts to exit, ubiquitous sound and light that provide continuous stimulation, and oxygen enrichment that maintains alertness despite sleep deprivation. These choices deliberately impair attentional control to maximize gambling duration. While extreme, this reveals how environmental design constitutes a form of power over cognitive states. Ethicist Shannon Vallor argues that technology and environmental designers bear responsibility for cultivating or corrupting human cognitive virtue. Applied to aesthetics, this suggests evaluating design choices not merely by efficiency or profitability metrics but by effects on human attentional flourishing, the capacity for sustained focus, contemplative depth, and meaningful engagement that constitute cognitive wellbeing.

The possibility of aesthetically designed environments that support rather than exploit attention appears in the architectural expressions of contemplative traditions. Cistercian monastery design eliminated decorative elements that might have drawn attention, creating spaces that support sustained focus on liturgical practice and meditative prayer. Paradoxically, this ascetic aesthetic produces beauty through proportion, natural light, and material honesty, qualities that support rather than distract from contemplative attention. Japanese tea ceremony architecture similarly removes ornamental excess while

39

achieving aesthetic refinement through careful attention to proportion, texture, and spatial sequence. These traditions suggest that the highest aesthetic achievement might involve not capturing attention through novelty and stimulation but creating conditions where attention rests naturally in present-moment awareness, unforced and undistracted.

Understanding aesthetics as attentional infrastructure rather than superficial decoration transforms design practice from styling to ethical intervention. Every chromatic choice, spatial proportion, material selection, and temporal rhythm either facilitates or obstructs the human cognitive capacities that constitute our highest possibilities: sustained focus enabling deep learning, contemplative awareness supporting wisdom development, and relational presence making authentic connection possible. The designed environments we create and inhabit either cultivate or corrupt these capacities, making aesthetic literacy an essential component of attentional self-determination in an increasingly developed world.

## Restoration Through Aesthetic Encounter

The capacity of aesthetic environments to restore depleted attentional resources reveals a reciprocal relationship between design and cognition that extends beyond mere stimulation or distraction. Environmental psychologists Rachel and Stephen Kaplan developed Attention Restoration Theory, proposing that natural environments provide four essential qualities supporting cognitive recovery: being away (psychological distance from routine demands), extent (sufficient richness to engage the mind without effort), soft fascination (stimuli capturing attention effortlessly), and compatibility (environments matching intrinsic inclinations). Urban parks demonstrate these principles in practice: Frederick Law Olmsted designed Central Park explicitly as a psychological refuge from Manhattan's sensory intensity, incorporating pastoral meadows, winding paths that reveal varied prospects, and water features that provide gentle acoustic masking. Contemporary research confirms his intuitions, even brief exposures to park environments measurably improve performance on tasks requiring sustained attention and working memory.

Built environments can incorporate restorative aesthetic principles through biophilic design strategies that reference natural patterns without requiring literal nature. Architect Michael Pawlyn's Sahara Forest Project greenhouse structures employ geometric patterns derived from radiolaria and diatoms, microscopic organisms whose forms optimize structural efficiency through evolutionary selection. These patterns satisfy the visual system's appetite for complexity within order, the same quality that makes natural environments restorative. Interior applications include living walls that introduce authentic vegetation into workspaces, circadian lighting systems that mimic diurnal changes in the solar spectrum, and water features that provide acoustic white noise to mask disruptive sounds without creating intelligible patterns that capture attention. The aesthetic sophistication lies in recognizing that humans evolved over millions of years in natural environments; our perceptual systems remain calibrated to these patterns, regardless of contemporary urban contexts.

The democratization of aesthetic design knowledge becomes crucial as pressure from the attention economy intensifies. Currently, a sophisticated understanding of how environmental aesthetics shape cognition remains concentrated among specialists serving elite clients or deployed by corporations maximizing engagement metrics. Educational initiatives teaching aesthetic literacy, how to recognize and evaluate the attentional affordances of designed environments, could empower individuals to make informed choices about the spaces they inhabit and to advocate for public environments that support rather than exploit attention. This represents not mere consumer education but the cultivation of civic capacity, recognizing that shared environmental aesthetics constitute common resources deserving protection on par with air and water quality.

# Chapter 4: Tech's Influence: Navigating Digital Attention

The smartphone you carry generates approximately 2.5 kilobytes of metadata every minute it remains active, including location coordinates, application usage patterns, screen-on duration, interaction sequences, and network connections. This continuous data exhaust, aggregated across billions of devices, feeds machine learning systems that predict with unsettling accuracy when you will next check your phone, which content will provoke the most extended engagement, and what emotional state will make you most susceptible to particular messaging. These prediction systems do not merely observe behavior; they actively shape it through optimization algorithms that treat human attention as a resource to be harvested with ever-increasing efficiency. The infrastructure coordinating this extraction operates largely invisibly to users, its complexity rivaling that of air traffic control systems or electrical grid management. Yet, its primary function involves not safety or utility but the systematic manipulation of consciousness for commercial purposes. Understanding technology's influence on attention requires examining not just the devices themselves but the economic imperatives, algorithmic logics, and interaction paradigms that transform human awareness into quantifiable, tradable assets.

The fundamental innovation enabling this transformation lies in the shift from information retrieval to information delivery systems. Early internet architecture required users to actively seek content through search queries, bookmark collections, and manually typed URLs, actions demanding intention and conscious choice. Contemporary platforms invert this relationship through recommendation algorithms that anticipate desires before users articulate them, serving content through feeds, suggestions, and notifications that require no active request. YouTube's recommendation engine, which accounts for over 70% of viewing time on the platform, employs collaborative filtering techniques that analyze billions of watch histories to surface videos that maximize session duration and drive return visits. The system optimizes neither for truth, educational value, nor user-stated preferences but for engagement metrics that correlate with advertising revenue. Research by sociologist Zeynep Tufekci documents how this creates "algorithmic rabbit holes", recommendation chains that lead from innocuous starting points to increasingly extreme content, exploiting the psychological tendency toward habituation, which requires progressively stronger stimuli to maintain interest. A user watching fitness videos may find suggestions gradually escalating to extreme dieting content, conspiracy theories about nutrition science, or communities promoting eating disorders, not through deliberate malice but through algorithmic optimization that has learned these trajectories maximize the metric it serves.

## The Neuroscience of Interface Addiction

The mechanisms by which digital technologies capture attention leverage specific vulnerabilities in dopamine-mediated reward pathways that evolved to motivate adaptive behaviors such as foraging, social bonding, and skill development. Variable-ratio reinforcement schedules, reward-delivery patterns in which reinforcement occurs unpredictably after varying numbers of responses, produce extraordinarily persistent behavior resistant to extinction. Neuroscientist Wolfram Schultz discovered that dopamine neurons fire most vigorously not when receiving rewards but during the anticipation of uncertain rewards, creating what psychologist Natasha Dow Schüll terms "the machine zone", a trance-like state in which awareness narrows to the interaction loop between action and an uncertain outcome. Slot machines exploit this through carefully calibrated near-miss outcomes that maintain player engagement despite net losses. Social media platforms implement parallel mechanics through unpredictable content quality in infinite scroll feeds, uncertain notification timing, and variable social feedback. When you refresh Instagram or Twitter, you cannot predict whether the new content will prove rewarding, creating the same anticipatory dopamine response that drives compulsive checking. Neuroimaging studies conducted by researchers at California State University reveal that chronic social media users exhibit neural activation patterns similar to those of substance-dependent individuals when anticipating platform access, including heightened amygdala activity and reduced prefrontal regulatory control.

The temporal structure of digital interactions fragments attention through what productivity researcher Gloria Mark terms "attention residue", the cognitive carryover from interrupted tasks that impairs performance on subsequent activities. Her empirical work tracking knowledge workers found that after being interrupted, people require an average of twenty-three minutes to return to their original task fully. However, they typically perceive the recovery time as only a few minutes. Yet the typical office worker now experiences interruptions every three to five minutes through emails, messages, notifications, and self-interruptions from checking devices. This creates a permanent state of divided attention, in which full cognitive engagement becomes impossible, regardless of willpower or intention. The mechanism involves working memory competition: when attention shifts to a new task before completing the previous one, mental resources remain allocated to the unfinished task, reducing capacity for the current task. Over time, this cognitive fragmentation becomes normalized, and workers lose the experiential baseline of sustained focus and may cease recognizing their own impairment. Mark's research revealed that workers consistently overestimate their multitasking efficiency while producing lower-quality output, making more errors, and experiencing elevated stress hormone levels compared to periods of continuous focus.

The persuasive technology industry has developed sophisticated A/B testing methodologies that enable rapid experimentation across millions of users simultaneously to identify interaction patterns that maximize engagement. Former Facebook engineer Jeff Seibert disclosed that the company routinely runs thousands of simultaneous experiments testing variations in interface design, content ranking, notification timing, and social feedback mechanisms. Each variation gets deployed to statistically significant user samples while metrics track resulting behavior changes at millisecond granularity. Features that increase session duration, return visits, or content sharing get progressively amplified through iterative refinement, while less "successful" variations disappear. This evolutionary process operates divorced from considerations of user well-being, privacy, or social consequences; the selection pressure is driven solely by commercial metrics. Particularly troubling, platforms systematically test emotional manipulation tactics: research published by Facebook's own data scientists described experiments deliberately skewing news feed content toward positive or negative emotional valence to measure resulting changes in user posting behavior. The study confirmed emotional contagion effects; manipulated users' subsequent posts reflected the emotional tone of content they had been shown, raising profound questions about consent, autonomy, and the ethical boundaries of corporate behavior modification.

## The Architecture of Distraction Economies

Understanding technology's grip on attention requires examining the business models of funding platform development. The attention economy operates through a three-party transaction structure fundamentally different from traditional commerce. Users provide attention and personal data without direct payment; platforms offer "free" services while collecting user information and controlling access; advertisers pay platforms for probabilistic access to user attention through targeted messaging. This arrangement creates perverse incentives where user interests systematically diverge from platform incentives. As technologist Jaron Lanier observes, when services are free, users become the product being sold. Platforms maximize revenue by increasing the quantity and quality of attention delivered to advertisers, creating inexorable pressure toward addictive design patterns, regardless of stated commitments to user well-being. The economic logic tolerates or even encourages practices that would constitute consumer protection violations in physical product markets: imagine pharmaceuticals optimized to maximize continued use regardless of health outcomes, or automobiles designed to discourage exiting once at destinations.

The advertising auction systems determining which content appears in user feeds introduce additional attention distortions. Google's AdWords platform pioneered real-time bidding, where advertisers compete for placement based on keyword relevance and willingness to pay, with prices determined through continuous auctions that handle billions of queries daily. While conceptually similar to traditional advertising markets, the granularity and targeting precision available through digital platforms create qualitatively different dynamics. Advertisers can target not just demographic categories but psychological profiles derived from behavioral tracking: individuals experiencing relationship difficulties, people searching for addiction treatment, and users exhibiting financial desperation. The informational asymmetry is stark; platforms possess detailed behavioral data on users, while users lack comparable information about advertisers or their persuasion strategies. Research by computer scientist Arvind Narayanan demonstrates that even "anonymized" browsing data can be re-identified with shocking reliability, enabling persistent tracking across devices and contexts that users believe are private. This surveillance infrastructure exists not primarily for security or functionality but to enhance advertising precision, treating human behavioral prediction as the core product being refined and sold.

The metrics governing platform success, daily active users, session duration, and revenue per user create optimization pressures extending beyond individual platforms into broader digital ecosystems. Application developers designing mobile software face economic incentives toward notification overuse, the creation of artificial urgency, and psychological manipulation because their commercial viability depends on winning attention share against thousands of competing applications. The result resembles a tragedy of the commons where individually rational design choices collectively degrade the attentional environment for everyone. Mobile gaming demonstrates this dynamic clearly: once-straightforward games evolved into elaborate psychological manipulation systems employing energy mechanics limiting play sessions (encouraging frequent check-ins), random reward systems (exploiting variable ratio reinforcement schedules), social obligation mechanics (leveraging reciprocity norms and fear of disappointing others), and artificial scarcity (creating time-limited opportunities generating urgency). The design language, initially developed for exploitative free-to-play games, has diffused into productivity applications, educational software, and social platforms. Increasingly, digital interactions employ these psychological levers regardless of domain appropriateness or user well-being.

## Algorithmic Curation and the Personalization Paradox

The recommendation algorithms curating digital experiences promise personalization, content tailored to individual preferences and interests. Yet this customization introduces subtle attention distortions that accumulate into social consequences. Mathematician Cathy O'Neil describes how algorithms encode their creators' priorities and biases into seemingly objective computational processes. YouTube's recommendation system optimizes for watch time without considering content veracity, educational value, or psychological health impacts. The algorithm remains agnostic to whether extended viewing stems from compelling scientific education or conspiracy theories that promote social distrust. In practice, emotionally arousing content systematically outperforms measured analysis in engagement metrics, creating selection pressure toward sensationalism regardless of accuracy. Political scientist Jonathan Albright's research mapping YouTube recommendation networks found that searches for political terms reliably led users toward increasingly partisan and extreme content, creating "walled gardens" of ideological conformity. The personalization ostensibly serving user preferences actually narrows informational diets, reducing exposure to challenging perspectives and nuanced positions that fail to generate strong engagement signals.

This algorithmic filtering operates primarily in an opaque manner, leaving users without visibility into why particular content appears while other material remains hidden. The curatorial power once held by human editors, librarians, and journalists, roles with professional standards, institutional accountability, and at least nominal public interest obligations, has transferred to proprietary algorithms optimized for commercial metrics and protected as trade secrets. When news feed algorithms determine that stories about distant famines warrant less prominence than celebrity gossip based on engagement predictions, they make editorial judgments with profound implications for public awareness and political priorities. Yet no democratic process governs these decisions, no public discourse shapes algorithmic values, and no accountability mechanism allows affected communities to contest their informational marginalization. Legal scholar Frank Pasquale describes this opacity as "the black box society," in which consequential decisions about information access, reputation, and opportunity are made through computational processes that resist external scrutiny.

The personalization paradox intensifies through filter bubbles, information ecosystems where algorithmic curation creates self-reinforcing cycles of ideological confirmation. Sociologist Eli Pariser coined the term to describe how personalized search results and social feeds can isolate users within their existing viewpoints, limiting exposure to contradictory information or alternative perspectives. While empirical research suggests filter bubbles may be less absolute than initially feared, most people encounter some ideological diversity online, the algorithmic amplification of preference-confirming content remains significant. Psychologist Jonathan Haidt's research on moral foundations reveals that political persuasion requires understanding opponents' values, which becomes impossible when algorithmic systems systematically hide opposing perspectives behind engagement-optimized content. The attention economy thus not only fragments individual focus but also fragments collective understanding, making democratic deliberation requiring shared factual foundations increasingly difficult to sustain.

## Dark Patterns and Manipulative Design

Beyond algorithmic content curation, interface design itself employs psychological manipulation tactics that technology ethicists term "dark patterns", design choices that trick users into actions contrary to their interests. Privacy researcher Harry Brignull catalogs these deceptive practices: confusing unsubscribe processes that require multiple clicks through deliberately unclear options; hidden costs revealed only late in checkout processes after users have invested time in selection; forced continuity where free trials automatically convert to paid subscriptions requiring active cancellation; disguised advertisements formatted to resemble editorial content; confirm shaming that employs guilt-inducing language for declining offers. These patterns exploit cognitive biases, time pressure, and interface complexity to manufacture consent for data collection, purchases, or continued engagement that users would refuse under clearer circumstances. While individual instances might seem trivial, their ubiquity creates an environment where constant vigilance is necessary to protect against manipulation, a state of defensive attention that precludes the trusting engagement required for genuine platform utility.

The deliberate exploitation of cognitive limitations extends to "infinite scroll" interfaces that eliminate natural stopping points. Traditional media possessed inherent boundaries: newspapers had final pages, television shows had endings, and magazines had back covers. These boundaries provided cognitive closure, allowing disengagement without the fear of missing subsequent content. Infinite scroll, employed across social media feeds, news sites, and content platforms, abolishes these boundaries deliberately to increase consumption. The technique combines several psychological mechanisms: removing decision points that might prompt disengagement; exploiting Zeigarnik effects to build tension around incomplete tasks; and creating intermittent reinforcement through variable content quality. Research by computer scientist Aza Raskin, who regrets inventing infinite scroll, estimates that the technique costs humanity approximately 200,000 human lifetimes annually through extended consumption that provides diminishing marginal utility. The design reflects an explicit prioritization of engagement metrics over user time value, treating human attention as an infinitely extractable resource rather than a finite resource deserving protection.

Notification systems represent perhaps the most aggressive attention capture mechanism, interrupting users wherever they are with demands for immediate engagement. The typical smartphone user receives between 63 and 80 notifications daily across multiple applications, each generating a micro-disruption that fragments continuous thought. Yet research by attention scholar Gloria Mark found that only 11% of notifications involve genuinely time-sensitive information requiring an immediate response; the vast majority serve purely to drive platform re-engagement. Applications deliberately employ notification strategies that exploit social obligations and the fear of missing out: social media platforms notify users about others' posts to create reciprocity pressure, e-commerce sites generate artificial urgency through inventory scarcity claims, and news applications frame routine updates as breaking developments that demand immediate attention. The cumulative effect transforms devices from tools serving user purposes into external attention directors, overriding personal priorities with commercial imperatives.

## Resistance Strategies and Technological Sovereignty

Recognizing technology's systematic attention exploitation creates pressure toward resistance strategies ranging from individual discipline to collective action. Digital wellness movements promote practices like disabling notifications, reducing display modes to grayscale to reduce visual appeal, enforcing application time limits to set usage boundaries, and scheduling "digital sabbaths" for complete disconnection. These individual tactics provide modest benefit. Research by psychologists at the University of Texas demonstrates that merely having smartphones visible, even when powered off, measurably reduces available cognitive capacity through "brain drain" effects, in which part of attention remains allocated to resisting device checking. Yet individual strategies face severe limitations when confronting corporate adversaries employing teams of psychologists, neuroscientists, and designers with real-time data on billions of users. The contest suggests that willpower and vigilance can protect against environmental toxins rather than demanding regulatory action to prevent pollution at the source.

More systemic approaches involve design movements toward "humane technology" championed by organizations like the Center for Humane Technology, founded by former technology insiders, including design ethicist Tristan Harris and investor Roger McNamee. These advocates promote design principles that respect user attention: bounded experiences with clear completion points, reduced interruption frequency, transparency about algorithmic curation, tools that enable user control over their informational environment, and business models aligned with user well-being rather than engagement maximization.

Some platforms have implemented partial reforms; Instagram now allows notification muting, YouTube provides watch-time dashboards, and Apple introduced Screen Time monitoring tools. However, these features coexist with core attention-harvesting mechanisms that remain unchanged, raising questions about whether voluntary industry reform can meaningfully address dynamics driven by competitive pressures and shareholder expectations demanding growth.

Legal and regulatory frameworks present alternative approaches through collective action, establishing minimum standards for corporate behavior. The European Union's General Data Protection Regulation represents the most comprehensive effort toward technological sovereignty, establishing user rights over personal data, requiring explicit consent for tracking, and imposing substantial penalties for violations. While focused primarily on privacy, these regulations indirectly constrain attention manipulation by limiting behavioral data collection, fueling algorithmic targeting. Some scholars advocate extending these principles to explicit "right to attention" protections: regulations requiring platforms to disclose engagement-optimization tactics, mandating user controls over algorithmic curation, prohibiting particularly manipulative dark patterns, or restricting advertising to contexts that do not exploit psychological vulnerabilities. Legal theorist Tim Wu proposes applying attention-protection frameworks analogous to environmental regulation, treating the collective attentional environment as a commons that requires protection from exploitative pollution, rather than leaving its defense to individual consumer choice.

Educational initiatives represent another response vector, though one facing significant challenges. Media literacy programs aim to develop critical awareness about technological manipulation tactics, algorithmic bias, and persuasive design. Yet effectiveness remains limited when adversaries possess sophisticated behavioral data while users rely on general principles and periodic training. Moreover, individual resistance strategies impose cognitive costs, maintaining constant vigilance against manipulation taxes the very attentional resources one seeks to protect. Philosopher Evan Selinger argues that this creates "privacy cynicism," in which people recognize manipulation but feel powerless to prevent it, leading to resignation rather than resistance. The dynamic parallels environmental pollution awareness without regulatory enforcement; knowledge alone is insufficient to protect against systemic harms that require collective action.

## Emerging Technologies and Attention Frontiers

Contemporary attention extraction systems may represent only preliminary stages of technological consciousness manipulation. Emerging technologies promise capabilities rendering current tactics crude by comparison. Virtual and augmented reality platforms create immersive environments with unprecedented sensory control, potentially intensifying both positive applications, such as educational simulations and therapeutic interventions, and exploitative possibilities. Neurotechnology companies develop consumer brain-computer interfaces ostensibly for communication and control, but these generate neurological data that could enable direct attention monitoring and manipulation. The ability to detect mental states, predict attention lapses, and identify decision-making moments from neural signals opens the door to persuasion operating beneath conscious awareness. Neuroethicist Nita Farahany warns that cognitive liberty, the right to private mental experience, faces unprecedented threats as technologies enable increasingly direct access to neural processes underlying attention, emotion, and decision-making.

As artificial intelligence capabilities advance toward artificial general intelligence, they raise profound questions about the autonomy of attention in human-machine interaction. Language models demonstrating increasingly sophisticated conversational abilities could enable personalized persuasion at scale, with AI agents adapting arguments in real time based on detected psychological vulnerabilities and resistance patterns. The development trajectory suggests systems that understand individual psychology more thoroughly than people understand themselves, creating asymmetries in which human attention is perpetually at a disadvantage against superhuman persuasion capabilities. Technology philosopher Shannon Vallor emphasizes that these developments demand proactive ethical frameworks rather than reactive responses after deployment. By the time attention manipulation capabilities become evident through harm, the technologies may have become so deeply embedded that removal proves economically or socially impossible.

Yet technology also enables resistance and the reclaiming of attentional sovereignty through tools that empower user agency. Open-source platforms, decentralized social networks, and privacy-focused alternatives demonstrate that commercial surveillance capitalism does not represent the only possible internet architecture. Projects like Mastodon, Signal, and DuckDuckGo prove technically viable alternatives to attention-harvesting platforms, though they struggle to achieve comparable adoption due to network effects favoring established services. The federation model, where independent servers interoperate while maintaining local control, opens the possibility of social connection without centralized manipulation. Similarly, algorithmic transparency initiatives and user-controlled recommendation systems allow individuals to understand and shape their informational environments rather than accepting opaque curatorial decisions optimized for commercial metrics rather than user benefit.

The fundamental question confronting digital attention concerns whether technological systems will serve human flourishing or exploit human vulnerability. Current trajectories suggest the latter absent significant changes in business models, regulatory frameworks, or collective social norms. The attention economy treats consciousness itself as raw material to be extracted rather than as the seat of human dignity deserving of protection and respect. Reclaiming attentional sovereignty requires moving beyond individual resistance tactics toward systemic transformations that address the economic incentives, technical architectures, and power asymmetries that enable systematic manipulation. Technology's influence on attention remains negotiable rather than inevitable, but negotiating successfully demands recognizing the magnitude of corporate investment in consciousness capture and responding with commensurate seriousness about protecting the cognitive foundations of autonomous human existence. The subsequent chapters will explore how this recognition translates into practical strategies for individuals and communities seeking to preserve attentional integrity within technological environments designed to subvert it.

# Chapter 5: Cultural Shifts: How Society Shapes Our Focus

The transformation of attention unfolds not merely through individual psychology or technological design but through vast cultural currents that redefine what societies collectively value, reward, and cultivate. Throughout human history, different civilizations have constructed radically divergent attentional ecologies, the social environments, practices, institutional structures, and shared assumptions that determine how communities allocate their most precious cognitive resource. Medieval monasteries organized entire lifetimes around sustained contemplation of sacred texts, with architectural silence and ritualized schedules protecting focus from worldly intrusion. Conversely, the Kwakiutl peoples of the Pacific Northwest developed potlatch ceremonies that demanded acute social attention, distributed across dozens of simultaneous gift exchanges, status negotiations, and reciprocity calculations. Neither culture possessed inherently superior attentional capacity; instead, each cultivated the specific forms of focus its social world required for meaningful participation. Understanding cultural influences on attention means recognizing that our current crisis emerges not from timeless human weakness but from historically specific contradictions between inherited attentional capacities and rapidly shifting cultural demands that have outpaced our collective adaptation mechanisms.

The Industrial Revolution inaugurated the first systematic attempt to standardize and commodify human attention through factory discipline. Before industrialization, most labor followed task-oriented rhythms, farmers worked according to seasonal necessities, artisans controlled their production tempo, and merchants determined their trading schedules. Historian E.P. Thompson documented how industrial capitalism required a fundamental reconstruction of temporal consciousness, transforming workers from task-oriented to time-disciplined subjects. Factory whistles, punch clocks, and foreman surveillance enforced unprecedented attentional conformity: workers must arrive simultaneously, maintain focus on repetitive operations for standardized durations, and subordinate personal rhythms to mechanical production schedules. This cultural revolution in attention management provoked intense resistance; early factory workers frequently abandoned employment during harvest seasons, celebrated "Saint Monday" with unauthorized absences, and engaged in workplace sabotage, reflecting deeper conflicts over who controlled the timing and focus of human activity. The eventual normalization of industrial time discipline, so complete that contemporary workers find pre-industrial work rhythms nearly incomprehensible, demonstrates how cultural transformation can fundamentally rewire collective attentional expectations within several generations.

## Education Systems as Attentional Training Grounds

The parallel rise of compulsory mass education during the nineteenth century created institutions explicitly designed to cultivate attentional dispositions compatible with industrial employment. Prussian educational reformers, whose model influenced systems worldwide, organized schooling around principles directly contradicting earlier pedagogical traditions. Where previous education for elites involved individualized instruction, Socratic dialogue, and self-directed study, mass schooling implemented standardized curricula, age-based cohorts, and synchronized classroom attention. Students learned to sit still for extended periods, redirect focus on command when bells signaled subject transitions, suppress personal interests in favor of assigned topics, and tolerate monotonous repetition, precisely the attentional skills factory employment required. Educational historian Joel Spring argues this represented cultural engineering rather than pedagogical necessity, training compliant workers rather than educated citizens. The legacy persists in contemporary schools, still organized around industrial-era assumptions: fifty-minute periods regardless of subject matter, simultaneous instruction despite diverse learning speeds, emphasis on attention compliance over genuine intellectual engagement, and assessment systems that reward sustained focus on externally imposed rather than intrinsically motivated tasks.

Contemporary educational reform movements reveal growing recognition that these inherited attentional structures serve neither students nor society effectively in post-industrial contexts. Finland's education system, frequently cited for superior outcomes, reduced instructional hours, eliminated standardized testing for younger students, increased recess frequency, and granted teachers substantial autonomy over pacing and methods. These changes implicitly recognize that attention flourishes in cultural conditions that differ from those that optimize factory discipline, conditions that prioritize intrinsic motivation, appropriate challenge levels, and respect for natural concentration rhythms rather than enforced uniformity. Similarly, Montessori education, now educating millions globally, structures learning environments around self-directed activity within prepared environments, allowing students to sustain focus on chosen tasks for extended periods rather than fragmenting attention across teacher-imposed transitions. Neuroscientific research by Angeline Lillard comparing Montessori and conventional students found significant advantages in executive function measures, including attentional control and task persistence, suggesting that cultural-educational structures that cultivate autonomy produce distinct attentional capacities compared to those that emphasize compliance.

The cultural dimension becomes particularly visible in cross-national comparisons of childhood experience. Anthropologist Elinor Ochs led longitudinal studies comparing Italian and American middle-class families, revealing stark differences in the socialization of attention. American parents constantly directed children's attention through questions, instructions, and educational interventions, "Look at the birdie," "What color is that?", treating children as projects requiring continuous adult guidance. Italian parents more frequently allowed children to direct their own attention, participating in adult activities peripherally until choosing to engage actively. These divergent practices reflect deeper cultural assumptions: American individualism emphasizes optimizing children's development through deliberate intervention, while Italian familism prioritizes social integration and self-regulated participation. The attentional consequences manifest in adolescence and adulthood, with Americans showing greater responsiveness to external direction but reduced capacity for sustained self-directed focus. At the same time, Italians demonstrate stronger intrinsic motivation but less immediate compliance with authority. Neither pattern proves universally superior; instead, each culture produces attentional dispositions adapted to its particular social arrangements, values, and economic structures.

## The Attention Recession and Economic Inequality

Contemporary capitalism has generated what economist James Williams terms an "attention recession", a systematic scarcity of focused awareness despite no reduction in absolute cognitive capacity, produced by cultural-economic structures that extract attention faster than individuals can regenerate it. Unlike industrial capitalism, which purchased workers' time, surveillance capitalism purchases the much more intimate resource of conscious experience itself. This represents a qualitative shift in what societies expect from human attention: not merely compliance during work hours but continuous availability for commercial targeting during all waking moments. Cultural critic Jonathan Crary identifies this trajectory toward "24/7 capitalism", a social order hostile to sleep, rest, and any attentional state resistant to monetization. The proliferation of "sleep optimization" advice, productivity systems for the evening hours, and pharmaceutical interventions that reduce sleep need all reflect cultural pressure toward total attentional availability. This pressure distributes unequally across social classes, creating what sociologist Judy Wajcman calls "time poverty" among low-income workers forced into multiple jobs with irregular schedules, offering neither the protected leisure time of pre-industrial societies nor the bounded work hours industrial unions secured.

The cultural valorization of "busyness" as a status marker inverts earlier class dynamics where leisure signified privilege. Sociologists Silvia Bellezza, Neeru Paharia, and Anat Keinan demonstrated through experimental studies that Americans perceive busy, overworked individuals as higher status than those with abundant free time, a complete reversal of Thorstein Veblen's "leisure class" dynamics from a century earlier. This cultural shift reflects economic restructuring: when wealth derives from knowledge work rather than property ownership, continuous mental availability signals importance and indispensability. The resulting "cult of overwork" creates social pressure to fragment attention across excessive commitments, with refusal interpreted as a lack of ambition rather than healthy boundaries. Silicon Valley's celebration of hundred-hour work weeks, Wall Street's expectation of weekend availability, and academia's normalization of constant productivity all reflect and reinforce this cultural transformation. The casualties appear in epidemic burnout rates, declining creative output despite longer hours, and the paradox of simultaneous information abundance and intellectual impoverishment.

Cultural attitudes toward attention deficit hyperactivity disorder reveal how societies pathologize attentional variations incompatible with dominant economic structures. ADHD diagnosis rates vary dramatically across nations and time periods in ways poorly explained by genetic differences. Psychiatrist Peter Breggin notes that American children receive ADHD diagnoses at rates multiple times higher than French children, reflecting not biological differences but contrasting cultural expectations about appropriate childhood behavior and divergent pharmaceutical industry influences. The behaviors diagnosed as disorders, difficulty sustaining attention on uninteresting tasks, preference for high-stimulation environments, and impulsivity would prove adaptive in many cultural contexts throughout history. Hunter-gatherer societies valued rapid attentional shifting to detect environmental changes; trading cultures rewarded quick decision-making and risk tolerance; artistic communities celebrated intense but episodic focus. The medicalization of these attentional styles reflects their incompatibility with contemporary schooling and office work rather than inherent pathology. Disability studies scholars argue this reveals how cultural structures disable people by demanding conformity to narrow attentional norms rather than accommodating human diversity, then classify non-conformers as individually defective rather than recognizing institutional failure.

## Media Revolutions and Collective Consciousness

Each major media revolution restructures cultural attention by transforming information accessibility, authority structures, and cognitive habits required for social participation. The transition from oral to literate cultures, extensively analyzed by classicist Walter Ong, fundamentally altered human consciousness by externalizing memory into written records. Oral cultures developed prodigious memorization capabilities, with individuals retaining vast poetic traditions, genealogies, and practical knowledge through techniques like rhythmic patterning and formulaic repetition. Literacy gradually atrophied these capacities while enabling new attentional modes: abstract reasoning divorced from immediate context, logical analysis of recorded arguments, and sustained focus on texts across multiple sessions. Plato's Phaedrus captures this transition's ambivalence. Socrates warns that writing will weaken memory and understanding, yet Plato records this warning in written dialogue, implicitly recognizing literacy's power despite its costs. The printing revolution further democratized access while standardizing texts, creating the possibility of the Protestant Reformation, individual Bible reading replacing clerical interpretation, and modern science, experimental results published for collective verification rather than guild secrets.

The telegraph initiated the compression of time and space in electric media, prompting what communications theorist James Carey termed the "divorce of transportation and communication." Previously, information traveled only as fast as physical messengers; telegraphy enabled instantaneous transmission across continents, fundamentally restructuring business, journalism, and military operations. This acceleration imposed new attentional demands: the stockbroker must monitor rapid price fluctuations, the journalist must produce "news" measured in hours rather than weeks, and the general must coordinate forces across vast distances through abbreviated messages. The resulting culture of speed and simultaneity, Marshall McLuhan's "global village", created both cosmopolitan awareness of distant events and attentional fragmentation as local engagement competed with global information flows. Subsequent media, radio, television, and the internet, intensified these patterns, each revolution provoking cultural anxiety about attentional decline. Socrates feared that writing would destroy memory; Victorians feared that novels would corrupt young women's morals through excessive imagination; twentieth-century critics feared that television would create passive, manipulable audiences. While specific fears proved exaggerated, each transition did restructure cultural attention in lasting ways, validating concerns about loss while underestimating adaptation and new capabilities emerging through altered practices.

Social media represents the most recent, and perhaps most profound, media revolution, restructuring attention at a cultural scale. Unlike broadcast media, where professionals produced content for mass audiences, social platforms position everyone simultaneously as producer and consumer, fundamentally democratizing but also fragmenting cultural discourse. The cultural consequences transcend individual platform effects examined in previous contexts. Anthropologist danah boyd documents how social media transformed adolescent identity formation by eliminating boundaries between previously distinct social contexts, family, school friends, romantic interests, and broader peer networks, that now collapse into a unified online presence. This demands continuous attention management across audiences with different expectations and norms, replacing earlier development periods where different social roles remained compartmentalized. The resulting "context collapse" generates anxiety and performativity, as authentic self-expression risks misinterpretation across diverse audiences. Additionally, social media's permanence and searchability eliminate cultural practices of forgetting that previously enabled social healing and personal reinvention. Youthful indiscretions, experimental identities, and contextual misjudgments persist indefinitely in digital archives, demanding constant attentional vigilance about potentially damaging historical traces.

## Ritual, Religion, and Collective Focus

Religious and ritual practices have historically served as primary cultural technologies for coordinating collective attention toward shared objects of significance. The Catholic Mass structured medieval European consciousness through daily bells that called communities to synchronized prayer, weekly liturgical cycles, and annual festivals that organized temporal experience around sacred narratives. This cultural architecture distributed attention across multiple timescales simultaneously: immediate ritual participation, daily rhythm, seasonal progression, and the eternal relationship with the divine. The Protestant Reformation's emphasis on individual Bible reading represented an attentional shift toward private textual engagement, contributing to rising literacy and the individualistic consciousness characterizing modernity. Judaism's Sabbath practice created weekly protected time explicitly refraining from productivity and commercial engagement, institutionalizing attentional rest as a religious obligation rather than a personal choice. Research by psychologist Andrew Newberg using neuroimaging during religious practices demonstrates that contemplative prayer, ritual chanting, and meditative states produce measurable changes in brain activity, including reduced parietal lobe activity associated with spatial orientation and self-other boundaries. These findings suggest religious practices functioned as cultural technologies for intentionally altering consciousness and attentional focus through reproducible methods.

The decline of religious participation in many industrial societies has not eliminated hunger for transcendent attentional experiences but dispersed them across substitute practices. Sociologist Émile Durkheim predicted that secular societies would develop "collective effervescence" through new institutions, such as sports spectatorship, political rallies, and music festivals, creating temporary unity through shared focus. Contemporary examples reveal both continuity and transformation: stadium crowds singing national anthems generate communal belonging comparable to religious congregations, yet remain episodic rather than woven into daily rhythms. The proliferation of "secular spirituality" through yoga, meditation apps, and wellness retreats represents explicit attempts to recover contemplative attentional practices stripped of religious metaphysics. However, the marketization of these practices, mindfulness as productivity enhancement, meditation as stress management, arguably domesticates their more radical potential to question the cultural structures that produce the attentional crisis in the first place. Scholar Jeff Wilson's analysis of "McMindfulness" critiques how corporate meditation programs teach employees to manage stress without questioning exploitative labor practices creating that stress, functioning as cultural pacification rather than genuine attentional liberation.

Indigenous cultures that maintain traditional practices offer alternative attentional models that resist complete subsumption within capitalist attention economies. Aboriginal Australian "walkabout" traditions involve extended periods away from settlements, navigating ancestral lands while maintaining attention to subtle environmental cues, songlines, and spiritual relationships with the landscape. Anthropologist Deborah Bird Rose describes this as "attentive consciousness", a mode of awareness that cultivates reciprocal relationships with the non-human world rather than treating nature as a resource for human exploitation. Similarly, many Indigenous American traditions include vision quests, extended fasting, and isolation practices specifically designed to alter ordinary consciousness and cultivate attention toward spiritual dimensions typically obscured by daily routines. These practices resist easy translation into dominant cultural frameworks precisely because they reject core assumptions about attention's proper objects and purposes. Where industrial culture treats attention as a resource for productive labor and consumption, these traditions position attention as a relationship with sacred dimensions requiring protection from profane intrusion. The survival of such practices demonstrates that alternative cultural arrangements remain possible despite overwhelming pressure toward global attentional standardization.

The proliferation of attention-related social movements signals growing cultural awareness that dominant attentional norms harm human flourishing. The "Slow Movement," originating with Carlo Petrini's Slow Food response to fast-food homogenization, expanded into Slow Cities, Slow Fashion, and a broader cultural critique of speed and efficiency as ultimate values. These movements don't merely advocate individual lifestyle changes but also challenge cultural structures that prioritize rapidity over quality, quantity over depth, and constant novelty over sustained engagement. Similarly, the "Right to Disconnect" legislation enacted in France, Spain, and other nations represents cultural recognition that employers' claims on employees' attention must be subject to legal limits rather than remain governed solely by economic power. These laws prohibit workplace communication outside contracted hours, institutionally protecting attentional recovery time against encroachment. Environmental philosopher Glenn Albrecht's concept of "solastalgia", distress caused by environmental change in one's home environment, extends to attentional environments: people experience genuine grief and disorientation as familiar attentional landscapes transform beyond recognition, replaced by alien ecologies hostile to inherited cognitive practices.

The emergence of digital minimalism, attention rebellion, and tech skepticism, particularly among younger generations, suggests potential cultural turning points. Unlike simple rejection, these movements demonstrate sophisticated engagement with technology's attentional effects while refusing inevitability narratives claiming resistance proves futile. Writer Jenny Odell's "How to Do Nothing" articulates principles for "resisting the attention economy" through strategic disengagement, not withdrawal from all digital interaction but careful curation of attentional commitments according to personally and politically meaningful priorities rather than algorithmic manipulation. The growing popularity of "dumb phones," digital sabbaths, and app-free zones indicates cultural experimentation with alternative relationships to technology beyond either wholesale adoption or complete refusal. These practices draw inspiration from multiple traditions: monastic discipline, Sabbath observance, Indigenous ceremony, counter-cultural resistance, assembling hybrid attentional cultures adapted to contemporary conditions while rejecting total subsumption within dominant economic logics.

Cultural transformation of attention operates simultaneously at multiple scales, from institutional policies to individual practices, from economic restructuring to symbolic meaning-making. No single intervention suffices; instead, sustainable change requires coordinated action across domains. Educational reform must align with labor market transformation, and developing intrinsic motivation proves futile if employment remains structured around compliance metrics. Similarly, individual attention management techniques provide only limited relief absent broader cultural shifts that constrain commercial exploitation of consciousness. The challenge facing contemporary societies involves nothing less than collective determination of what human attention ultimately serves: Will it remain primarily an economic resource extracted for private profit, or can cultures reassert attention's role in relationship, creativity, contemplation, and the construction of meaningful lives? The answer will be written not through individual choices alone but through cultural movements, institutional reforms, and political struggles, determining what kind of attentional environment humanity collectively constructs for itself and future generations. Cultural forces shape focus not through deterministic imposition but through structured possibilities, the options made available, the practices rewarded, and the alternatives rendered visible or obscured. Understanding these cultural dynamics represents the essential foundation for anyone seeking not merely to manage their own attention more effectively but to participate in transforming the broader attentional ecology within which all contemporary consciousness unfolds.

# Chapter 6: The Neuroscience Behind Concentration

The human brain operates through approximately eighty-six billion neurons, each forming thousands of synaptic connections that create networks of staggering complexity. Yet concentration, the sustained, voluntary directing of mental resources toward a single target, depends not on the brain's totality but on the coordinated activity of specific neural circuits that have evolved to prioritize some information while actively suppressing competing inputs. This selective amplification occurs through oscillatory synchronization, whereby neurons firing in coordinated rhythms create what neuroscientist Pascal Fries terms "communication through coherence." When disparate brain regions oscillate at matching frequencies, particularly in the gamma band (30-100 Hz), they establish functional connectivity that enables information transfer between distant cortical areas. During intense mental effort, the prefrontal cortex synchronizes with posterior sensory regions, forming temporary coalitions that bind attention to perceptual targets. Simultaneously, theta oscillations between 4 and 8 Hz appear in hippocampal and prefrontal areas, providing the temporal scaffolding that sequences working memory operations. The neuroscience of concentration thus reveals not a single "attention center" but distributed systems that achieve temporary unity through rhythmic coordination, a neural democracy in which voting occurs through synchronized firing patterns rather than anatomical hierarchy.

The anterior cingulate cortex, nestled in the medial wall of the frontal lobes, functions as an executive monitoring system that detects conflicts between competing response tendencies and signals the need for increased cognitive control. Neuroscientist Michael Petrides demonstrated, through lesion studies, that damage to this region impairs the ability to maintain concentration in the face of distractions, though basic attention remains intact. The mechanism involves continuous error monitoring; the anterior cingulate compares intended actions against actual performance, generating error-related negativity signals within eighty milliseconds of mistakes. This rapid feedback allows real-time adjustments that keep behavior aligned with goals despite interference. During sustained concentration tasks, functional magnetic resonance imaging reveals anterior cingulate activation proportional to conflict demands, suggesting this region titrates control according to need rather than maintaining constant vigilance. Chronic stress, however, alters this calibration by sustained cortisol elevation, which reduces anterior cingulate gray matter density. Research led by Rajita Sinha at Yale School of Medicine found that individuals experiencing prolonged stress show both structural thinning in this region and functional impairments in conflict monitoring, manifesting as reduced concentration capacity when facing competing demands. The finding illuminates why periods of intense stress often coincide with inability to focus, not from lack of motivation but from compromised neural architecture underlying conflict detection and resolution.

## The Noradrenergic System and Attentional Optimization

Concentration depends critically on optimal arousal levels, mediated by the locus coeruleus, a small brainstem nucleus containing merely 15,000 neurons that nevertheless projects throughout the entire cortex, releasing norepinephrine that modulates neural gain, the responsiveness of neurons to their inputs. Too little norepinephrine produces drowsiness and inattention; excessive release generates anxiety and distractibility. Neuroscientist Gary Aston-Jones discovered that the locus coeruleus operates in two distinct modes: tonic firing produces steady baseline norepinephrine levels, maintaining general alertness. At the same time, phasic bursts occur in response to task-relevant stimuli, temporarily enhancing sensory processing and behavioral responding. The optimal concentration state involves moderate tonic activity and robust phasic reactions, resulting in what Aston-Jones terms "engaged task performance." However, when challenges intensify or persist for too long, the locus coeruleus shifts toward purely tonic, high-frequency firing that indiscriminately floods the cortex with norepinephrine, destroying the signal-to-noise ratio that enables selective attention. This transition explains the concentration failures that accompany acute stress or exhaustion; the system meant to enhance focus becomes dysfunctional through overactivation, creating distractibility precisely when focus is most crucial.

The Yerkes-Dodson principle, formulated a century before neuroscience could explain its mechanisms, captured this inverted-U relationship between arousal and performance: moderate arousal optimizes concentration, while both understimulation and overstimulation impair it. Contemporary research reveals the neurochemical substrates underlying this behavioral law. Neuroscientist Amy Arnsten's laboratory demonstrated that moderate levels of norepinephrine and dopamine enhance prefrontal cortex function by strengthening relevant neural signals through postsynaptic receptor actions. In contrast, excessive catecholamine release impairs prefrontal function through alternative receptor pathways that trigger signal collapse. This explains individual differences in optimal concentration conditions: morning larks possess different baseline arousal patterns than night owls, introverts maintain higher basal arousal than extroverts, and anxious individuals show elevated noradrenergic tone requiring less environmental stimulation for optimal focus. Personalizing concentration strategies, therefore, requires understanding one's neurochemical starting point rather than applying universal recommendations that assume identical baseline states across individuals.

## Acetylcholine and the Precision of Focused Attention

While norepinephrine broadly regulates arousal, the neurotransmitter acetylcholine provides more targeted attentional enhancement through projections from the basal forebrain to sensory cortices. Neuroscientist Michael Hasselmo's computational models demonstrate that acetylcholine alters cortical dynamics by suppressing internal associative processing while enhancing responses to external sensory input. This neurochemical shift enables what might be termed "reception mode", a brain state optimized for encoding new information rather than elaborating existing knowledge. During concentrated learning, acetylcholine release in sensory cortices sharpens receptive field tuning, making neurons more selective for their preferred stimuli while reducing responses to marginal features. This creates heightened perceptual discrimination, allowing detailed attention to subtle distinctions that would typically fall below awareness thresholds. Research by Yuka Minces, using optogenetic techniques to stimulate cholinergic neurons artificially, demonstrated that even in the absence of external tasks, acetylcholine release alone improves sensory discrimination and accelerates learning rates in experimental animals.

The relationship between acetylcholine and concentration becomes pathologically disrupted in Alzheimer's disease, where degeneration of basal forebrain cholinergic neurons produces catastrophic attentional deficits long before memory failure dominates the clinical picture. Early-stage patients exhibit profound difficulties sustaining concentration during conversations, reading, or routine tasks despite preserved intelligence and motivation. Pharmaceutical interventions using acetylcholinesterase inhibitors, drugs that prevent acetylcholine breakdown, thereby increasing synaptic availability, produce modest but meaningful improvements in attentional capacity, supporting the link between cholinergic function and concentration. However, these medications reveal that simply elevating acetylcholine cannot restore normal concentration, suggesting that the precise temporal patterning of release matters as much as absolute levels. Natural acetylcholine dynamics involve rapid phasic changes that are synchronized with attentional shifts, whereas pharmacological approaches produce only tonic elevations, lacking temporal specificity. The therapeutic limitations highlight how concentration depends not merely on neurochemical presence but on precisely orchestrated release patterns coordinated with cognitive demands.

## Neural Plasticity and the Trainability of Concentration

The capacity for concentration demonstrates remarkable plasticity across the lifespan through experience-dependent neural reorganization. London taxi drivers, who spend years memorizing the city's twenty-five thousand streets to earn licensure, show enlarged posterior hippocampi compared to matched controls, not from genetic predisposition but from sustained spatial learning demands. Neuroscientist Eleanor Maguire discovered that hippocampal volume correlated with time spent navigating, demonstrating that intensive concentration on particular cognitive domains literally reshapes brain structure. Subsequent research revealed that these changes involve not only neuronal growth but also modifications in white matter tracts connecting the hippocampus with prefrontal and parietal regions, which are essential for spatial attention and memory integration. The structural adaptations persist years after active navigation ceases, suggesting that sustained concentration produces enduring architectural changes rather than temporary functional adjustments.

Meditation practices provide another window into concentration-induced neuroplasticity through interventions explicitly designed to strengthen attentional control. Neuroscientist Richard Davidson's laboratory conducted long-term studies with Tibetan Buddhist monks who had accumulated over 10,000 hours of meditation practice, finding extraordinary gamma-band oscillatory activity during focused-attention meditation that far exceeded that observed in novices. More remarkably, randomized controlled trials with meditation-naive adults showed that even eight weeks of daily practice produced measurable changes in brain structure and function. Sara Lazar's neuroimaging studies revealed cortical thickening in the dorsolateral prefrontal cortex and anterior insula, regions critical for attentional control and interoceptive awareness, following mindfulness training. Plasticity occurs through multiple mechanisms, including synaptogenesis, dendritic branching, myelination, and angiogenesis, which collectively enhance neural processing efficiency in trained circuits. Behavioral improvements accompany these structural changes: meditators show reduced attentional blink magnitude, faster attentional disengagement from distractors, and superior performance on vigilance tasks requiring sustained concentration across extended periods.

The temporal dynamics of concentration-related plasticity follow distinct phases corresponding to different neurobiological processes. Initial practice sessions produce rapid functional changes through synaptic modifications that strengthen relevant neural pathways within hours, explaining why single training sessions can yield immediate, though fragile, performance improvements. Consolidation occurs over subsequent days through protein synthesis and structural changes at synapses, transforming transient enhancements into stable modifications that resist decay. Finally, extended practice over months triggers systems-level reorganization, including white matter changes that enhance communication efficiency between distant brain regions. This multiphase trajectory means that concentration training requires both immediate practice benefits and long-term commitment to develop substantial capacity. Weekend workshops or brief interventions can introduce techniques and demonstrate possibilities, but enduring concentration enhancement emerges only from sustained practice, enabling progressive neural reorganization across successive plasticity phases.

## The Default Mode Network and Concentration Failures

Concentration does not simply involve activating attentional systems. Still, it requires simultaneous suppression of the default mode network, a constellation of brain regions including the medial prefrontal cortex, the posterior cingulate cortex, and the angular gyrus that becomes active during rest and mind-wandering. Marcus Raichle discovered this network through neuroimaging studies that revealed surprising patterns of brain activity during passive baseline conditions between experimental tasks. Rather than simply idling, the resting brain engages in stimulus-independent thought, mental time travel, social cognition, and self-referential processing that consumes substantial metabolic resources despite producing no external behavior. During demanding concentration tasks, default network activity decreases proportionally to task difficulty, suggesting active suppression rather than passive deactivation. The reciprocal relationship between task-positive networks supporting concentration and the default mode network creates what neurologist Marcus Raichle terms an "anti-correlated" pattern. As one system activates, the other necessarily suppresses.

Concentration failures often reflect insufficient default network suppression rather than inadequate attentional activation. Neuroscientist Jonathan Smallwood's experience-sampling studies, which combined subjective reports with neuroimaging, demonstrated that mind-wandering episodes correlate with heightened default network activity intruding into task-positive states. During these momentary lapses, attention detaches from external tasks and drifts toward internal concerns, worries about upcoming events, rumination about past interactions, or spontaneous memory retrieval. The neurological substrate involves a brief imbalance in which default network activity escapes prefrontal inhibitory control, capturing processing resources that are usually allocated to external concentration. Individual differences in working memory capacity predict susceptibility to these lapses; people with superior working memory maintain more stable task focus because they possess greater resources for simultaneously managing task performance and default network suppression. Conversely, reduced working memory increases vulnerability to internal distraction, as insufficient resources remain for active inhibition of the default network.

Interestingly, complete default network suppression proves neither possible nor desirable for optimal cognitive functioning. Creativity research by psychologist Scott Barry Kaufman demonstrates that insight and novel problem-solving often emerge during moments when attention relaxes from concentrated focus, allowing default network activity to make remote associations and retrieve relevant but not explicitly sought information. The challenge involves not eliminating default network activity but achieving an appropriate dynamic balance, concentrated focus when tasks demand it, and relaxed, unfocused states when they facilitate creative synthesis. Pathological concentration appears in obsessive-compulsive disorder, where individuals become locked into task-positive states, unable to disengage, producing perseveration and cognitive inflexibility. The optimal pattern involves fluid transitions between concentrated external focus and relaxed internal reflection, with prefrontal executive systems coordinating these shifts according to situational demands rather than getting stuck in either mode.

## Neurotransmitter Interactions and Concentration Dynamics

Concentration arises from the dynamic interplay among multiple neurotransmitter systems rather than from any single neurochemical agent. While norepinephrine regulates arousal and acetylcholine tunes sensory processing, dopamine confers motivational salience, determining what captures sustained attention. Wolfram Schultz's groundbreaking work revealed that dopamine neurons encode reward prediction errors, the difference between expected and received outcomes, thereby teaching the brain which environmental features merit attentional investment. When outcomes exceed predictions, dopamine bursts signal that attended features possess greater value than anticipated, strengthening attentional orienting toward those cues. Conversely, when outcomes fall short, dopamine dips below baseline, weakening associations between attention and disappointing targets. This learning mechanism explains why concentration naturally flows toward intrinsically rewarding activities. Dopamine reinforcement has strengthened neural pathways connecting those activities with sustained attentional engagement.

The dopaminergic system's sensitivity to prediction errors, while adaptive for learning, creates vulnerability to exploitation through artificially manipulated reward schedules. Variable reinforcement ratios produce sustained dopaminergic responding because outcomes remain perpetually uncertain, generating prediction errors even without net reward increases. This neurochemical quirk underlies the concentration capture by digital platforms, leading to unpredictable content quality and social feedback. Each notification, message, or feed refresh potentially contains rewarding information, maintaining dopamine-driven attentional orienting even when average outcomes prove disappointing. Neuroscientist Robert Sapolsky demonstrated that this creates a dissociation between wanting and liking, dopamine drives motivated attention toward uncertain rewards without necessarily producing pleasurable experiences upon delivery. The result resembles addiction's core features: compulsive attention toward specific stimuli despite reduced enjoyment and awareness that behavior patterns undermine long-term interests.

Serotonergic systems modulate concentration through effects on patience and temporal discounting, the degree to which delayed rewards lose subjective value. Neuroscientist Molly Crockett's research using acute tryptophan depletion, which temporarily reduces serotonin synthesis, demonstrated that lower serotonergic tone increases impulsivity and preference for immediate over delayed gratification. Applied to concentration, this suggests that serotonin supports sustained focus on long-term goals by reducing the allure of immediate alternatives. Individuals with chronically low serotonin show difficulty maintaining concentration on effortful tasks because moment-to-moment competing options, checking phones, switching activities, and seeking immediate stimulation, become disproportionately attractive compared to larger future rewards requiring sustained present effort. The serotonergic system thus provides the neurochemical foundation for what psychologist Walter Mischel termed delay of gratification, enabling concentration to persist despite attractive alternatives constantly beckoning for attentional diversion.

## The Metabolic Demands of Sustained Concentration

The brain constitutes merely 2% of body mass, yet it consumes approximately 20% of metabolic energy, with concentrated mental effort demanding disproportionate resources beyond baseline neural activity. Neuroscientist Marcus Raichle calculated that focused cognitive tasks increase regional brain metabolism by only five to ten percent above resting levels, a surprisingly modest increment given the subjective sense of exertion accompanying difficult concentration. However, this small percentage translates into substantial absolute energy demands because resting brain metabolism already operates near the physiological maximum sustainable rate. Concentrated effort pushes local brain regions toward metabolic ceilings limited by glucose delivery, oxygen availability, and waste product clearance through cerebral blood flow. Neurovascular coupling mechanisms increase blood flow to active areas, but these compensatory responses require seconds to engage fully, creating temporary energy deficits during initial concentration demands.

The metabolic costs accumulate during sustained concentration through depletion of local neurotransmitter pools, accumulation of adenosine (a byproduct of ATP breakdown that promotes drowsiness), and buildup of other fatigue-related metabolites. Neuroscientist Nilli Lavie demonstrated that cognitive control, the executive processes maintaining concentration against distraction, proves especially metabolically expensive, consuming disproportionate resources relative to automatic processing. This explains the subjective sense of exhaustion following periods of intense focus, even in the absence of physical activity: the brain has depleted local energy reserves and accumulated metabolic waste products that need to be cleared. Recovery periods allow replenishment through glucose uptake and waste removal, but these restorative processes require time proportional to the duration of depletion. Attempting to sustain concentration beyond metabolic capacity produces diminishing returns as neural efficiency degrades, reaction times lengthen, error rates increase, and subjective effort feelings intensify while actual performance deteriorates.

Individual metabolic efficiency varies substantially based on genetic factors, fitness levels, and nutritional status, creating differences in sustainable concentration duration that cannot be overcome through willpower alone. Aerobic fitness enhances cerebral blood flow and metabolic efficiency, explaining why cardiorespiratory conditioning improves cognitive endurance independent of muscular benefits. Blood glucose regulation affects concentration stability through insulin sensitivity, which determines how effectively neurons access circulating glucose. Individuals with insulin resistance experience concentration difficulties because, despite adequate blood glucose levels, cellular uptake mechanisms function poorly, leading to functional hypoglycemia at the neural level. The metabolic perspective reframes concentration not as a character virtue but as a physiological capacity constrained by energetic limitations, suggesting that sustainable focus requires respecting biological constraints rather than attempting to override them through determination alone.

The neuroscience of concentration ultimately reveals a system of extraordinary sophistication operating through principles fundamentally different from those suggested by conscious experience. Rather than a unitary attention spotlight controlled by willful intention, concentration emerges from distributed neural networks that coordinate through oscillatory synchronization, modulated by multiple neurotransmitter systems each contributing distinct functional properties, constrained by metabolic limitations, and shaped by experience-dependent plasticity. This mechanistic understanding simultaneously humbles and empowers, humbles by revealing how concentration depends on neurobiological machinery operating primarily outside conscious control, yet empowers by identifying specific mechanisms amenable to intervention through training, environmental modification, and physiological optimization. The subsequent chapters will build on this neurobiological foundation to explore practical applications for cultivating concentration capacity within the constraints and possibilities afforded by brain architecture.

# Chapter 7: Mindful Practices: Cultivating Deep Attention

The ancient monastery bell rings at four in the morning, not to impose arbitrary discipline but to harness a specific window when the prefrontal cortex emerges from sleep with maximal plasticity for attentional training. This precise timing reflects centuries of empirical observation by contemplative traditions that, through lived experience, discovered what neuroscience now confirms: attention is not a fixed trait but a cultivable skill that responds to systematic practice with measurable improvements in stability, selectivity, and metacognitive awareness. The practices emerging from Buddhist, Stoic, Sufi, and indigenous contemplative lineages share a common recognition that the untrained mind resembles a wild horse, reactive and undirected. In contrast, the cultivated mind becomes responsive without losing spontaneity. Contemporary attention science validates these insights while revealing the specific mechanisms through which deliberate practice restructures neural architecture, supporting sustained focus. Understanding these practices requires moving beyond the superficial association of mindfulness with relaxation or stress relief to examine how specific techniques target distinct attentional capacities through graded challenges that progressively expand cognitive control.

The systematic cultivation of attention begins with developing metacognitive monitoring, the capacity to observe one's own mental states without immediate identification or reaction. Traditional Vipassana meditation employs this through body-scan practices, in which practitioners systematically move awareness through physical sensations from feet to head, noticing whatever sensations arise, tingling, pressure, temperature, pain, while maintaining an observational distance rather than evaluating them as pleasant or unpleasant. This seemingly simple practice engages multiple sophisticated cognitive mechanisms simultaneously. The deliberate movement of attention from one body region to another strengthens voluntary attentional control circuits involving the dorsolateral prefrontal cortex and the posterior parietal areas. The sustained maintenance of awareness within each location exercises concentration stability mediated by thalamic gating that filters competing sensory inputs. The non-judgmental observation cultivates what psychologist Phillippa Lally terms "response flexibility", creating temporal space between stimulus and reaction where conscious choice becomes possible rather than reflexive reactivity dominating behavior. Neuroimaging studies by Judson Brewer at Brown University demonstrate that experienced meditators show reduced activity in the posterior cingulate cortex during practice, indicating quieting of self-referential processing that usually colors perception with egocentric evaluation. This neural signature corresponds phenomenologically to the subjective sense of observing sensations without immediately relating them to personal narratives of suffering or satisfaction.

## Concentration and Open Awareness: Complementary Attentional Capacities

Mindful practices systematically develop two complementary attentional modes that together constitute comprehensive cognitive flexibility. Focused attention meditation, exemplified by breath awareness practices, trains the capacity to maintain continuous contact with a single object despite inevitable distractions. Practitioners anchor awareness on breath sensations at the nostrils or abdomen, noting when attention wanders and gently returning it without self-criticism. This apparent simplicity conceals profound training demands; the typical beginning practitioner discovers attention strays within seconds, revealing how little voluntary control exists over supposedly voluntary consciousness. However, longitudinal studies by Clifford Saron, tracking meditation retreat participants over 3 months, revealed systematic improvements in attentional stability, as measured by behavioral tasks requiring sustained vigilance and by EEG markers of attention-related processing. The improvements persisted for five months post-retreat, demonstrating lasting changes rather than temporary state effects during practice sessions. The mechanism involves progressive strengthening of endogenous attention networks that generate top-down biasing signals, allowing practitioners to sustain focus on intended objects against the constant competition from salient distractors that automatically capture awareness in untrained individuals.

Open monitoring meditation develops the complementary capacity for panoramic awareness without selective focus, maintaining receptive consciousness of whatever arises in experience without grasping after pleasant contents or pushing away unpleasant ones. Zen shikantaza ("just sitting") exemplifies this approach: practitioners maintain upright posture and open eyes, allowing thoughts, sensations, emotions, and perceptions to arise and pass without engaging with or suppressing them. Unlike focused attention, which involves effortful concentration on a chosen object, open monitoring cultivates effortless presence with the entire field of experience. This practice targets different neural systems; rather than strengthening top-down control mechanisms, it reduces interference from emotional reactivity and conceptual elaboration that typically hijack awareness. Research by Wendy Hasenkamp at Emory University using real-time fMRI during open awareness meditation found distinctive patterns involving decreased activation in narrative self-referential regions combined with heightened sensitivity in sensory processing areas, suggesting practitioners achieve stable conscious registration of sensory information while reducing conceptual overlay that typically filters immediate experience through memory, anticipation, and judgment.

The integration of these complementary capacities produces attentional agility, the ability to fluidly shift between focused concentration when tasks demand it and receptive awareness when circumstances require monitoring multiple information streams without predetermined priorities. Traditional meditation training sequences this development carefully: the initial months emphasize concentration practices that build basic stability before introducing open monitoring, recognizing that panoramic awareness without sufficient concentration degenerates into scattered distraction rather than clear presence. The sequential training parallels how elite athletes develop specific skills through isolated drills before integrating them into fluid performance during competition. Cognitive scientist Shinzen Young developed mathematical frameworks quantifying this integration through what he terms "sensory clarity, concentration power, and equanimity", three factors that interact multiplicatively rather than additively. Doubling concentration with minimal clarity produces only incremental improvement, but cultivating both together creates exponential gains in functional attention capacity applicable to demanding real-world contexts beyond formal practice periods.

## Anchoring Practices: Working with Physical Sensations

The breath serves as the primary anchor in countless contemplative traditions precisely because it provides continuous sensory feedback that bridges voluntary and involuntary control systems. Unlike arbitrary objects that require sustained visual focus and rapidly produce eye strain, or mantras that remain purely mental constructs, breathing offers tangible sensations accessible in any posture or environment and continues automatically when attention wanders. This makes breath ideal for developing what meditation teacher Joseph Goldstein describes as "continuity of mindfulness", the capacity to maintain unbroken awareness through changing conditions rather than concentration achievable only in optimal circumstances. Advanced practices employ increasingly subtle breath-related sensations as objects, progressively refining perceptual discrimination. Initial practice focuses on gross sensations of breath movement, chest or abdominal expansion, and contraction. As stability develops, attention narrows to sensations at the nostril rim where air contacts skin during inhalation and exhalation. Eventually, practitioners detect subtle temperature gradients between incoming and outgoing breath, or the momentary pause between inhalation and exhalation, where breathing temporarily ceases. This graduated refinement parallels visual acuity training, where athletes learn to track faster objects or musicians distinguish increasingly subtle pitch variations, the perceptual system becomes calibrated to detect signal characteristics previously lost in noise.

Alternative somatic anchors offer distinct training benefits for different attentional challenges. Walking meditation, performed at an unnaturally slow pace where each footstep unfolds over five to ten seconds, develops awareness of complex movement sequences typically executed unconsciously through cerebellar automation. Practitioners deconstruct each step into component phases, lifting the heel, shifting weight, moving the foot forward, lowering the foot, contacting the ground, shifting weight onto the stepping foot, and maintaining continuous awareness through the entire cycle before initiating the next step. This practice targets proprioceptive and kinesthetic senses rather than exteroceptive senses like vision or hearing, building comprehensive sensory awareness that is not limited to a single modality. Neurologist Frank Wilson's research on hand-brain interaction revealed that deliberate attention to movement increases the density of motor cortex representations, expanding neural territory devoted to those movements and enhancing control precision. Walking meditation exploits this neuroplasticity to recover conscious access to automated movement patterns, creating fluidity between conscious intention and embodied action that reduces the dissociation between mental activity and physical presence that characterizes typical distracted awareness.

## Cultivating Metacognitive Precision Through Mental Noting

Mental noting practices, particularly emphasized in Mahasi Sayadaw's Burmese Vipassana tradition, employ linguistic labels to sharpen discrimination between mental phenomena that normally blur together in unreflective experience. Practitioners apply brief mental labels to whatever becomes predominant in awareness, "thinking," "hearing," "tingling," "planning," "remembering", using labels not to suppress or judge experiences but to clearly recognize their nature before returning attention to primary objects. This technique addresses a fundamental challenge in attention training: the mind resists observing itself because the observer becomes entangled with observed contents, lacking the separation necessary for clear perception.

Labeling creates cognitive distance, allowing practitioners to recognize mental activities as events occurring within awareness rather than defining awareness itself. The practice cultivates what philosopher Thomas Metzinger terms "epistemic agency", the capacity to recognize that thoughts constitute mental constructs rather than direct representations of reality, thereby fostering appropriate skepticism about the reliability of unexamined mental content.

The specificity of noting develops progressively from broad categories toward increasingly fine discriminations. Beginning practitioners might use only three labels, "thinking," "feeling," and "sensing", to distinguish cognitive, emotional, and perceptual experiences. As precision increases, thinking subdivides into planning, remembering, fantasizing, analyzing, or worrying. Emotions can manifest as frustration, disappointment, anxiety, contentment, boredom, or impatience. Physical sensations resolve into pressure, temperature, tingling, pulsing, aching, or itching. This granular awareness serves multiple functions beyond meditation practice itself. Research by Lisa Feldman Barrett at Northeastern University demonstrates that emotional granularity, the ability to generate precise emotion labels, correlates with superior emotion regulation and reduced reactivity to stressful situations. The mechanism involves constructing more differentiated internal models that enable nuanced responses rather than crude categorizations that trigger stereotyped reactions. Someone who distinguishes between disappointment and resentment can respond appropriately to each, while someone experiencing only undifferentiated "upset" deploys the same maladaptive strategies regardless of specific circumstances. Mental noting thus builds the conceptual infrastructure supporting sophisticated emotional intelligence and behavioral flexibility that extend far beyond formal practice contexts.

Advanced noting practices target the temporal microstructure of experience, revealing that apparently continuous mental activity actually consists of discrete momentary events that arise and pass with extraordinary rapidity. Traditional texts describe experiencing consciousness as a series of instantaneous knowing events rather than a continuous stream, likening it to how film creates the illusion of motion from discrete still frames. Contemporary practitioners employing high-frequency noting, labeling mental events as rapidly as possible, sometimes multiple times per second, report profound shifts in temporal perception, in which the typically opaque flow of consciousness becomes transparent, revealing the constructed nature of subjective continuity. Neuroscientist Francisco Varela's neurophenomenological research program attempted to bridge this first-person experiential precision with third-person neurophysiological measurement, suggesting that contemplative training provides the disciplined introspection necessary for rigorous consciousness science. While methodological challenges limit definitive conclusions, the approach acknowledges that cultivated attention offers investigative capacities unavailable through untrained subjective reports or purely objective measurement, potentially revealing aspects of consciousness inaccessible through conventional neuroscience methods.

## Integration Practices: From Cushion to Chaos

The ultimate test of cultivated attention involves maintaining stability and clarity while navigating demanding environments rather than only during protected practice periods. This integration challenge represents the crucial difference between attention as a recreational hobby versus a transformative capacity reshaping lived experience. Traditional contemplative training emphasizes gradual environmental progression; initial practice occurs in quiet, controlled settings with minimal sensory stimulation. As stability develops, practitioners intentionally introduce complexity: practicing with ambient noise, in uncomfortable temperatures, during illness, or amid emotional turbulence. Eventually, formal practice boundaries dissolve as practitioners attempt to maintain continuous mindfulness through daily activities, eating, cleaning, conversing, working, and treating every moment as an opportunity for attention training rather than dividing life into practice periods and ordinary experience.

The kitchen provides an ideal laboratory for integrating mindful attention into complex activities requiring multiple cognitive capacities simultaneously. Cooking demands sustained focus on recipe steps while monitoring numerous parallel processes, simmering sauces that require occasional stirring, timer-dependent operations like baking, and sequential preparations where earlier steps must complete before later ones begin. Bringing contemplative attention to cooking transforms routine necessity into a comprehensive training ground. Practitioners attend fully to sensory details typically processed automatically, the specific sound of onions hitting hot oil, the precise moment when garlic fragrance shifts from raw to aromatic to beginning bitter, the resistance of the knife blade through vegetables, revealing the sharpness quality. This sensory richness provides continuous anchors, preventing mind-wandering while complex task demands exercise cognitive control capacities beyond simple breath meditation. Professional chef and meditation teacher Edward Espe Brown describes cooking as "Zen practice in action," noting that attentive cooking cultivates presence, patience, and responsiveness while producing tangible results that provide immediate feedback on attention quality. Burned food reveals distraction, while well-executed dishes reflect sustained focus.

Interpersonal interaction represents perhaps the most challenging integration context because social situations involve unpredictable dynamics that trigger emotional reactivity, destabilizing attention more powerfully than environmental stimuli. Contemplative listening practices cultivate attention specifically adapted to conversational demands. Practitioners commit to maintaining complete presence with speakers without planning responses, judging content, or drifting into personal associations triggered by topics mentioned. This requires suspending the regular conversational pattern in which attention alternates between brief listening and formulating replies, creating dialogue that is more like a series of monologues than a genuine exchange. Psychologist Carl Rogers termed this "unconditional positive regard", maintaining an accepting presence regardless of the message's content, separating the person from the positions they express. Research on therapeutic alliance demonstrates that this quality of attention predicts positive outcomes across diverse therapy modalities more reliably than the specific techniques employed, suggesting that sustained, non-judgmental presence itself creates conditions that facilitate change. The practice proves extraordinarily difficult precisely because conversations activate deeply conditioned response patterns, the urge to share similar experiences, offer advice, defend against perceived criticism, or redirect discussion toward preferred topics. Maintaining receptive attention against these impulses requires the same mental muscles developed in breath meditation, now applied in the far more complex and emotionally charged interpersonal domain.

## The Science of Practice Parameters: Duration, Frequency, and Intensity

Optimizing attention cultivation requires understanding how practice parameters affect developmental trajectories. The common assumption that longer practice sessions invariably produce superior results is incorrect; research reveals complex relationships between quantity and quality, with intermediate dosages often outperforming both minimal and maximal extremes. Studies comparing meditation practice effects across different durations found that twenty to thirty-minute daily sessions produce more consistent benefits than either ten-minute abbreviated sessions or hour-long extended sessions for most practitioners. The mechanism involves balancing sufficient duration for meaningful engagement against fatigue that degrades attention quality. Brief sessions prevent reaching states where concentration stabilizes and insights emerge, while excessive duration exhausts cognitive resources, transforming practice into an endurance ordeal rather than skilled training. The optimal window appears when practitioners achieve stable focus without undue effort, maintaining quality awareness until natural fatigue signals appropriate stopping points rather than forcing continuation through willpower that produces tension and aversion toward future practice.

Frequency proves more critical than duration for establishing lasting changes in attention capacity. Daily practice, even briefly, outperforms intensive weekend sessions followed by gaps in establishing stable neural reorganization. The principle mirrors physical training: consistent moderate exercise produces superior fitness adaptations than sporadic intense workouts followed by recovery periods. Neuroscientist Alvaro Pascual-Leone's research on motor skill acquisition using transcranial magnetic stimulation demonstrated that daily practice leads to progressive expansion of motor cortex representations. In contrast, practice every third day produces minimal change despite equivalent total practice time. Continuous daily practice appears necessary to trigger molecular cascades involving brain-derived neurotrophic factor and synaptic plasticity mechanisms that consolidate new functional architectures. In practice, this suggests that establishing a non-negotiable daily practice, however brief, should precede attempts at intensive retreats or extended sessions. The practitioner who maintains twenty minutes daily builds a more robust attention capacity than one who completes occasional weekend workshops, despite greater absolute practice hours in the latter case.

Intensity parameters involve not duration but the degree of mental effort applied during practice. Beginning practitioners often approach meditation with intense striving, attempting to force concentration through sheer willpower and becoming frustrated when attention wanders. This counterproductive tension actually impairs attentional development by activating stress responses that fragment rather than stabilize awareness. Advanced practitioners describe optimal practice involving "relaxed effort" or "effortless concentration", maintaining clear intention without forceful striving, combining alertness with ease. Meditation teacher Shinzen Young uses the metaphor of holding a bird: grip too loosely and it escapes; grip too tightly and you crush it; the precise, light firmness allows secure holding without harm. Translating this phenomenological description into a trainable skill requires graduated difficulty progression. Initial practice should employ easy conditions, comfortable posture, minimal ambient stimulation, short durations before fatigue arises, allowing practitioners to discover what stable attention feels like without excessive effort. As stability develops, practitioners gradually introduce challenges, longer sessions, less ideal environments, and more distracting circumstances that require increased effort to maintain quality awareness. This progressive overload principle, familiar from physical training, prevents both insufficient challenge, which does not adapt, and excessive challenge, which creates discouragement and technique breakdown.

The prospect of cultivating attention through systematic practice offers liberation from the assumption that focus constitutes an innate trait distributed unequally and unchangeably across individuals. While genetic factors and early developmental experiences certainly influence baseline attentional capacity, the neuroplasticity research reviewed throughout this chapter demonstrates that deliberate training produces meaningful improvements regardless of starting point. The practices described here represent only a subset of diverse contemplative traditions that, over centuries of empirical refinement, have discovered how to develop attention as a skill systematically. Their integration into contemporary life need not require adopting entire cultural or religious frameworks; the techniques prove effective when understood as cognitive training protocols targeting specific neural systems supporting attentional control. As technology and culture continue conspiring to fragment focus, these proven methods for cultivating deep attention become not spiritual luxuries but practical necessities for anyone seeking to reclaim sovereignty over their own consciousness.

# Chapter 8: Attention and Identity: The Self in the Spotlight

The photograph you choose as your profile image initiates a cascade of cognitive processes more consequential than the simple act suggests. In selecting that particular angle, that specific expression, that carefully curated background, you make decisions about which version of yourself deserves public attention, and by extension, which aspects of your identity warrant existence. Psychologist Dan McAdams describes identity as a "life story" we construct and revise continuously, but the attention economy has transformed this narrative process into something far more literal and immediate. Every social media post, every shared article, every commented response broadcasts identity claims to audiences whose attention validates or invalidates the self-concepts being projected. The contemporary self increasingly exists as a performance that requires constant attentional investment, monitoring how we appear, managing impressions, tracking responses, and adjusting presentations. This transforms identity from an internal integration of experience into an external construction dependent on captured attention. When philosopher Simone de Beauvoir wrote that "one is not born, but rather becomes, a woman," she identified how identity emerges through social recognition. The digital age amplifies this dynamic exponentially: we become whoever can successfully command attention in algorithmically mediated social spaces, regardless of whether that performed self corresponds to internal experience or authentic values.

The reciprocal relationship between attention and identity operates through what sociologist Erving Goffman termed "impression management", the strategic control of information to influence how others perceive us. Goffman's dramaturgical framework, developed through observing face-to-face interactions in the 1950s, distinguished between "front stage" behavior performed for audiences and "backstage" regions where roles could be dropped. Digital life collapses this boundary: the smartphone accompanies us backstage, transforming every moment into potential performance. Developmental psychologist Erik Erikson identified adolescence as the crucial period for identity formation, when young people explore different roles and commitments to forge coherent self-concepts. Contemporary adolescents conduct this exploration in unprecedented conditions; every experimental identity exists in searchable, permanent digital records; every social misstep circulates through peer networks at algorithmic speed; every provisional self-presentation generates quantified feedback through likes, shares, and comments. Research by psychologist Jean Twenge analyzing cross-generational data sets reveals marked increases in self-monitoring and impression-management concerns among individuals reaching adolescence after 2010, coinciding with the ubiquity of smartphones. The correlation suggests that constant attentional availability reshapes identity development itself, creating selves oriented primarily toward external validation rather than internal coherence.

## Attentional Currency and the Quantified Self

The transformation of personal attributes into metrics fundamentally alters how identity forms and stabilizes. Philosophers from Aristotle through William James understood identity development as the cultivation of character through habitual action; we become virtuous through practicing virtue, courageous through acting courageously, and thoughtful through thinking carefully. This classical model locates identity formation in the quality of attention we direct toward activities and values rather than the attention we receive from others. The quantified self movement inverts this directionality: identity becomes defined through accumulated metrics, steps walked, calories consumed, followers gained, and engagement rates achieved. These numbers provide concrete feedback loops replacing the ambiguous internal signals that previously guided self-development. Philosopher Byung-Chul Han argues this produces what he terms "the achievement-subject," whose identity derives entirely from measurable outputs and comparative rankings rather than intrinsic qualities or relationships. The fitness tracker displaying daily movement statistics creates an identity as "someone who walks 10,000 steps," regardless of whether walking produces genuine satisfaction or merely serves as a metric to optimize.

This quantification extends beyond physical activities into previously unmeasurable domains of personality and social value. The Klout score purported to represent social influence numerically; dating applications reduce complex human compatibility to algorithmic match percentages; professional networking platforms transform career accomplishments into profile-completeness meters. Each metric creates an attentional focal point that shapes self-concept: we become concerned with attributes the measurement highlights, while dimensions excluded from quantification recede in relevance to identity. Anthropologist Natasha Dow Schüll, whose research initially examined slot machine gambling, later investigated how self-tracking technologies create similar compulsive monitoring behaviors. The parallel emerges through uncertainty: just as gamblers cannot predict when machines will pay out, self-trackers cannot know when their metrics will show desired improvements, creating persistent checking behaviors that fragment attention across multiple quantified identity dimensions. The resulting self resembles an investment portfolio requiring constant monitoring rather than an integrated person living from internalized values.

The pressure to achieve quantifiable identity performance particularly affects domains traditionally considered intrinsic to authentic selfhood. Creative production, once valued for its expressive or aesthetic dimensions, is now primarily evaluated through attention metrics, views, downloads, and shares. Musicians describe the experience of creating songs optimized for streaming algorithms rather than artistic vision, knowing that Spotify's recommendation system determines discoverability based on engagement patterns within the first 30 seconds—writers structure articles around SEO keywords that improve search visibility rather than arguments that matter to them intellectually. The identity "creative artist" transforms from someone producing meaningful work into someone successfully gaming attention allocation systems. Media theorist Wendy Chun identifies this as a "crisis of liveness," in which authentic present experience becomes impossible because every moment exists primarily as content for future attentional capture. The photographer experiencing a sunset finds the experience mediated by considerations of how the scene will photograph, which filters will optimize visual impact, and which caption will generate engagement. Attention is directed entirely toward future audience reception rather than present sensory immersion.

## Parasocial Attention and Vicarious Identity

The relationship between attention and identity extends beyond personal performance into identification with figures who successfully command mass attention. Parasocial relationships, one-sided attachments in which audiences develop strong connections to media personalities who remain unaware of individual fans, now constitute a primary source of identity formation for substantial segments of the population. Psychologist Donald Horton first identified this phenomenon in 1956, observing television viewers and noting how people incorporated favorite characters into their social worlds despite the fundamental asymmetry of the relationship. Contemporary parasocial bonds operate through radically different mechanisms enabled by interactive platforms: influencers respond to comments (occasionally), appear in ephemeral "stories" that mimic intimate communication, and share personal details to create authenticity impressions. Research by social psychologist Jennifer Escalas demonstrates that narrative transportation, becoming absorbed in characters' experiences, facilitates vicarious identity adoption. Followers begin defining themselves through their relationships with admired figures: "I'm a Taylor Swift fan" becomes not merely a taste preference but a core identity dimension that organizes social connections, consumption choices, and self-understanding.

This identification process intensifies through what communication scholars call "illusory intimacy," where carefully constructed accessibility signals mislead audiences into perceiving genuine mutual relationships. The influencer's first-person address ("Hey guys!"), Morning routine documentation and vulnerability displays ("struggling today, needed to share this") employ intimacy's formal properties while maintaining commercial relationships fundamentally incompatible with authentic friendship. Philosopher Judith Butler's work on performative identity emphasized how repeated actions create the illusion of pre-existing essential selves; influencers extend this principle industrially, performing intimacy so consistently that audiences mistake the performance for reality. The attentional economy incentivizes this confusion: influencers monetize relationships through sponsored content, affiliate commissions, and platform revenue-sharing, creating structural pressure to maximize perceived closeness while serving thousands or millions of simultaneous "relationships." Followers orient their attention toward these figures with such intensity that parasocial bonds can overwhelm face-to-face connections. Teenagers prioritize following influencer updates over conversing with physically present family members, not because of antisocial personality traits, but because algorithmic platforms have successfully convinced them that these mediated relationships constitute their genuine social world.

The identity implications become particularly concerning when parasocial attention focuses on figures explicitly monetizing aspiration gaps. The Instagram model whose income derives from promoting weight-loss products profits directly from followers who feel inadequate about their bodies. The luxury lifestyle influencer's brand partnerships depend on the audience's desire for unattainable levels of consumption. The productivity guru's course sales require followers to believe their current lives lack sufficient optimization. These relationships create identity feedback loops where attention directed toward aspirational figures generates self-dissatisfaction that drives continued engagement seeking solutions the influencer conveniently sells. Psychologist Shira Gabriel's research on parasocial relationships reveals that they can provide genuine psychological benefits, reduce loneliness, increase self-esteem, and feelings of community belonging. Still, these benefits correlate specifically with relationships to figures who display authentic vulnerability and mutual value expression rather than those that promote commercial aspiration. The distinction proves critical: parasocial attention can support healthy identity development when it exposes people to diverse perspectives and role models demonstrating values-aligned living, but becomes pathological when it creates a perpetual sense of inadequacy that requires constant consumption to resolve.

## Attention Scarcity and Identity Fragmentation

The proliferation of platforms demanding attentional presence across different contexts creates what sociologist Sherry Turkle identifies as "identity fatigue," in which the cognitive load of maintaining multiple performed selves across various audiences produces exhaustion and alienation. Pre-digital life allowed considerable audience segregation: professional colleagues never encountered your weekend recreational activities; childhood friends remained separate from romantic partners; family saw you differently than your artistic community did. This compartmentalization enabled complexity; different facets of multidimensional selves could coexist without requiring integration into unified, coherent identities. Social media's network convergence forces these separate audiences into shared spaces, creating what researcher danah boyd calls "context collapse," in which the duplicate content is broadcast to parents, employers, romantic interests, and acquaintances simultaneously.

This collapse generates two contrasting adaptation strategies, each with distinct identity implications. Some individuals perform sanitized, generic selves acceptable across all possible audiences, eliminating anything potentially controversial, personally revealing, or contextually specific. This produces what philosopher Charles Taylor termed "flattened selfhood," in which rich internal complexity is reduced to bland, lowest-common-denominator presentations. The performed identity becomes detached from genuine experience, creating dissociative splits between public persona and private reality. Alternatively, people create separate accounts for different audiences, professional LinkedIn, personal Instagram, and anonymous Twitter, fragmenting identity across disconnected platforms. This strategy allows greater authenticity within each context but requires substantial cognitive effort to track which self occupies which space, remember where information exists, and prevent inadvertent context bleeding. Both adaptations impose significant attentional costs: constant self-monitoring to ensure appropriate performance or mental partitioning to maintain separate identity streams.

The fragmentation extends temporally through permanent digital records that freeze past identity versions in searchable public archives. Developmental psychology demonstrates that healthy maturation requires revision of earlier self-concepts as people gain experience, develop values, and integrate new perspectives. The "right to be forgotten" legal frameworks emerging in European jurisdictions recognize this developmental necessity by allowing the deletion of outdated personal information. However, algorithmic memory proves far more tenacious than human memory: search results surface decades-old content; social media companies preserve deleted posts in backups; archive services capture website snapshots. The teenage self who posted inflammatory opinions remains perpetually accessible, potentially defining adult identity despite subsequent growth. Writer Emily Nussbaum describes this as "life with receipts," where every past action and statement remains available as evidence. The resulting hyperaccountability creates conservative pressure against identity experimentation, knowing that provisional positions become permanent records discourages the exploratory testing essential for mature self-understanding. Young people increasingly avoid committing to positions or affiliations, not from indecisiveness but from rational assessment that premature commitment creates future constraints as searchable history defines their identity independently of actual current values.

## Attentional Deprivation and the Invisible Self

While excessive attention creates performed, surveilled selves, attention scarcity generates equally profound identity distortions. Sociologist Orlando Patterson's concept of "social death" described how enslaved people were systematically denied recognition as persons with legitimate claims to attention, dignity, or self-determination. Contemporary marginalized communities experience analogous dynamics through structural attention deprivation; their experiences, perspectives, and concerns remain systematically excluded from spaces where social reality gets constructed. Critical race theorist Ralph Ellison's Invisible Man literalized this phenomenon: the protagonist remains unseen not through physical transparency but through others' refusal to acknowledge his humanity, looking through rather than at him. This attentional erasure doesn't merely ignore specific identities; it prevents their formation by denying the recognition that Hegel identified as essential to the development of self-consciousness. We know ourselves through others' recognition; when recognition systematically fails, coherent identity becomes profoundly difficult to establish.

The attention economy exacerbates these dynamics through algorithms that determine visibility based on engagement metrics that favor dominant-group perspectives. Research by communication scholar Safiya Noble documents how search engine algorithms systematically surfaced racist imagery and stereotypes in results for identity-related queries, not through explicit programming but through optimization toward click-through rates that reflected broader societal biases. The Black teenager searching for images of "professional hairstyles" encounters results implicitly suggesting their natural hair violates professional norms. This attentional pattern shapes identity by constantly signaling which versions of self warrant recognition versus suppression. Disability activist Alice Wong describes how social media platforms structurally disadvantage disability communities through accessibility failures, missing image descriptions, incompatible screen readers, autoplay videos causing seizures, and creating digital spaces where disabled identities literally cannot participate. The resulting attention monopoly by non-disabled users doesn't simply exclude disability perspectives; it constructs abled identity as an unmarked default while rendering disability invisible and therefore seemingly non-existent as a valid identity category.

Language itself encodes attention hierarchies that determine which identities achieve articulation and which remain unnamed and therefore conceptually unavailable. Linguist Lera Boroditsky's research demonstrates how grammatical structures shape thought. Languages that require gender marking for all nouns make speakers more likely to attribute gender characteristics to inanimate objects. Applied to identity, communities lacking a vocabulary for particular identity configurations find those identities difficult to recognize or claim. The recent proliferation of terminology around gender and sexual identity, nonbinary, genderfluid, asexual, aromantic, demisexual, provides linguistic resources enabling people to identify experiences previously lacking names. Each term creates attentional categories that legitimize identities previously forced into inappropriate binary classifications. Philosopher Miranda Fricker's concept of "epistemic injustice" describes how marginalized groups suffer hermeneutical disadvantage when dominant conceptual frameworks lack terms for their experiences, rendering specific identities literally unintelligible rather than merely stigmatized.

The consequences compound across generations as attention structures shape which identity possibilities children perceive as available. The absence of media representation for particular identity configurations, LGBTQ+ characters, disabled protagonists, and neurodivergent heroes creates "symbolic annihilation," where specific ways of being remain outside imaginable options. Media representation studies consistently demonstrate that children's vocational aspirations, gender role concepts, and identity possibilities expand dramatically when they encounter diverse representation in media commanding cultural attention. Conversely, populations systematically excluded from attention-commanding narratives internalize that exclusion as evidence of their lesser legitimacy. The cycle perpetuates: attention deprivation prevents identity articulation, which reinforces exclusion from attention-getting spaces, further limiting the possibilities for identity development.

## Reconstructing Authentic Identity Through Attentional Agency

Breaking free from attention economy identity distortions requires cultivating what philosopher Miranda Anderson calls "attentional sovereignty", reclaiming deliberate control over where attention flows and what recognition sources matter for self-understanding. This involves rejecting the premise that identity requires external validation through attention metrics, instead grounding self-concept in values, relationships, and activities chosen for intrinsic rather than performative significance. Existentialist philosopher Søren Kierkegaard distinguished between aesthetic existence, living for external approval and novelty, and ethical existence, choosing commitments that define authentic selfhood regardless of social recognition. Translating this to a contemporary context means directing attention toward developing capabilities and relationships that constitute genuinely meaningful life rather than optimizing for visibility. The carpenter refining woodworking technique through private practice, the gardener cultivating plant knowledge through seasons of attentive observation, the friend maintaining correspondence through handwritten letters, each resists attention economy logic by investing cognitive resources in activities that produce no measurable social capital yet develop rich identity grounding.

This reconstruction requires deliberate curation of attentional inputs that shape identity formation. Media literacy advocate Renee Hobbs emphasizes "reflective media diets," in which individuals consciously select information sources that support their values rather than passively consume algorithmic recommendations. Applied to identity, this means actively seeking recognition from sources that reflect authentic self-concepts rather than accepting validation only from attention economy platforms. Joining communities organized around shared practices, amateur astronomy clubs, running groups, and book discussion circles provides identity-relevant recognition based on substantive engagement rather than performed impressions. Psychologist Margarita Azmitia's research on identity development demonstrates that adolescents who form identity through community participation in valued activities develop more stable, resilient self-concepts than those who rely primarily on peer approval. The mechanism involves internalization: repeatedly receiving recognition for particular qualities or contributions creates self-concepts that incorporate those attributes as defining features, independent of continued external validation.

Therapeutic interventions increasingly recognize attention patterns as central to identity pathology and recovery. Acceptance and Commitment Therapy, developed by psychologist Steven Hayes, explicitly targets psychological flexibility, the capacity to direct attention toward chosen values despite uncomfortable thoughts or feelings. Rather than attempting to eliminate negative self-concepts, the approach trains clients to notice when attention becomes captured by self-critical narratives and deliberately redirect it toward values-aligned action. This creates identity through attentional commitment: repeatedly choosing to see and act on particular values gradually constructs selves defined by those values, regardless of cognitive content. Narrative therapy, pioneered by Michael White, works with clients to identify and elaborate preferred identity stories, versions of self that emphasize agency, values, and meaningful relationships rather than problem-saturated narratives that emphasize pathology. The process involves systematically directing clients' attention toward experiences that support preferred identities while reducing attention to experiences that reinforce problematic self-concepts. This doesn't deny difficulties but changes their definitional role: problems become challenges the person faces rather than defining characteristics determining identity.

The possibility of authentic identity in attention economies ultimately depends on cultivating what philosopher Iris Murdoch termed "unselfing", the capacity to direct attention genuinely toward others and the world rather than constantly monitoring how one appears. Murdoch argued that moral development requires escaping the ego's gravitational pull through disciplined attention toward reality independent of self-interest. Applied to identity, this suggests that paradoxically, the most coherent selves emerge when attention flows primarily outward toward valued activities, relationships, and purposes rather than inward toward self-construction. The musician absorbed in musical expression rather than audience impression, the scientist fascinated by experimental questions rather than career advancement, the parent attentive to children's developmental needs rather than parenting performance, each discovers identity as a byproduct of absorbed engagement rather than as a primary project requiring constant attention. This doesn't mean ignoring how others perceive us or abandoning legitimate needs for recognition. Instead, it involves rebalancing attention allocation: external recognition becomes helpful feedback rather than the foundation of identity; self-awareness serves navigation rather than constant self-monitoring. The self that emerges proves more stable precisely because it depends less on captured attention and more on sustained commitment to what genuinely matters beyond performance metrics.

# Chapter 9: The Economics of Attention: Value in the Modern Market

When Warren Buffett describes his investment philosophy, he emphasizes the concept of "economic moats", sustainable competitive advantages protecting companies from rivals. In the twenty-first century, the most valuable moats no longer consist of factories, patents, or distribution networks, but instead of captured human attention. Apple's market capitalization exceeding three trillion dollars reflects not merely the physical devices it manufactures but the ecosystem of habitual attention these devices command. Users check their iPhones an average of 96 times daily, creating billions of micro-moments when consciousness becomes available for monetization through app purchases, service subscriptions, and advertising exposure. This represents a fundamental transformation in value creation; economic worth increasingly derives from the capacity to attract, retain, and direct conscious awareness rather than from tangible asset ownership. Understanding attention as a financial resource requires examining how markets now price consciousness itself, how attention scarcity creates winner-take-all dynamics, and how this commodification fundamentally restructures social relationships beyond purely commercial contexts.

Traditional economic theory distinguishes between rival goods, in which one person's consumption prevents another's use, and non-rival goods, such as knowledge, which can be used by multiple people simultaneously. Attention occupies a unique category: it remains absolutely rival at the individual level; a person cannot simultaneously focus on various objects, yet aggregated attention across populations enables network effects that make platforms exponentially more valuable as user bases expand. Facebook demonstrates

111

this principle: while each user possesses strictly limited attention, the platform's value grows superlinearly with membership because social connection opportunities multiply with network size. Metcalfe's Law suggests that network value increases proportionally to the square of the number of connected users, but attention networks exhibit even steeper scaling through algorithmic amplification. A viral video reaching one million viewers creates attention cascades in which subsequent viewers arrive partly because others have already watched, leading to self-reinforcing engagement spirals. This dynamic produces what economist Hal Varian terms "attention leverage"; minor initial differences in attention are amplified by positive feedback, leading to tiny quality distinctions yielding massive outcome disparities. The tenth-best smartphone might offer 95% of the leading device's functionality but capture only 5% of market attention, because attention begets attention through social proof mechanisms that make popular choices appear inherently superior regardless of objective quality differences.

The financialization of attention has created derivative markets in which consciousness is traded multiple steps removed from actual human experience. Programmatic advertising exchanges execute billions of real-time auctions daily, purchasing not specific ad placements but probabilistic access to particular demographic attention during microsecond windows as web pages load. Attention futures emerged from bookings for advertising slots during events like the Super Bowl, where companies commit millions for 30-second segments years in advance, speculating on future attention availability and engagement rates. More abstractly, venture capital valuations of pre-revenue startups fundamentally represent attention speculation; investors bet that nascent platforms will eventually capture sufficient user consciousness to justify extraordinary present valuations through future monetization. Instagram's one-billion-dollar acquisition by Facebook in 2012, when the service generated zero revenue, was priced purely based on the attention the platform commanded and the threat it posed to Facebook's attention monopoly. This speculative attention valuation creates profound market distortions: companies that optimize for growth metrics signaling attention capture receive funding regardless of profitability or social value, while businesses that generate actual value but attract limited attention struggle to attract capital.

## The Productivity Paradox and Attention Misallocation

Economist Robert Solow observed in 1987 that "you can see the computer age everywhere but in the productivity statistics," identifying a paradox where massive technological investment produced disappointing economic output gains. Contemporary analysis reveals that attention misallocation substantially explains this puzzle. Knowledge workers spend an estimated 28% of their work time managing email, not because email enhances productivity, but because organizational attention norms mandate constant availability. Studies by economist Gloria Mark tracking information workers through typical days found that they spend only 11 minutes on any project before interruption, and that it takes 23 minutes to return to the original task. This fragmentation destroys deep work capacity, the sustained, focused attention needed for complex problem-solving, strategic thinking, and creative synthesis. While information technology theoretically enhances worker capabilities, the attention costs imposed through constant connectivity and communication overhead eclipse potential productivity gains. Economist Erik Brynjolfsson terms this the "productivity J-curve," where new technologies initially reduce measured productivity during adoption periods as organizations adapt workflows and train users. However, attention technologies may represent a fundamentally different case: rather than temporary adjustment costs, they impose permanent attention taxes that cannot be eliminated through better adaptation because attention fragmentation serves the economic interests of technology providers even when harming user productivity.

This dynamic manifests most clearly in what economists call "negative externalities", costs imposed on third parties who did not choose to participate in the transaction. When a company adopts communication platforms that require employees' constant availability, the attention costs fall on workers, whose lives become subordinated to perpetual responsiveness. At the same time, productivity gains accrue to the organization. Workplace surveillance technologies are marketed as productivity tools, keystroke monitors, website tracking, and attention analytics, which extract maximum attention from workers while creating massive psychological costs through diminished autonomy and trust. The economic logic treats human attention as infinitely renewable, subject to unlimited extraction without depletion effects, contradicting both psychological research on cognitive fatigue and basic thermodynamic principles regarding limited metabolic resources supporting brain function. Economist William Nordhaus, analyzing long-term growth trends, argued that technological progress primarily manifests as increased consumer surplus, value captured by users rather than producers. Attention technologies reverse this pattern: producers capture the majority of value through advertising revenue and data monetization, while users experience net negative welfare through reduced well-being, privacy loss, and attention degradation that exceeds the utility provided by ostensibly free services.

The labor market implications extend beyond productivity into wage dynamics and inequality. Economist David Autor's research on labor-market polarization found that middle-skill jobs requiring moderate training and concentration have declined precipitously. In contrast, both high-skill analytical positions and low-skill service jobs expanded. This hollowing out partly reflects attention economics: routine cognitive tasks became automatable precisely because they required only moderate sustained attention, while work demanding either exceptional focus or emotional attentiveness remains difficult to mechanize. However, the attention economy creates additional stratification through differential ability to command market attention itself. Knowledge workers whose output depends on attention from others, consultants, analysts, and teachers, find their economic value increasingly tied to personal brand building and visibility cultivation rather than technical competence. Time spent on self-promotion through professional networking, content creation, and engagement farming constitutes uncompensated labor required to remain economically viable, effectively lowering real wages while creating winner-take-all dynamics where the most visible practitioners capture disproportionate rewards. Economist Sherwin Rosen described similar effects in entertainment and sports as "superstar economics," but attention markets universalize this pattern across previously stable professions.

## Attention: Accounting and Hidden Economic Costs

Conventional economic measurement systematically understates attention costs because they are not reflected on traditional balance sheets. Companies report physical capital investments, buildings, equipment, infrastructure, but treat employee attention as a free input requiring no maintenance or regeneration. This creates perverse accounting in which technologies imposing massive attention costs are recorded as pure efficiency gains. Email, by eliminating secretarial labor and postal costs, appears to be a productivity enhancement despite the hidden costs of perpetual interruption and cognitive task-switching that devastate concentrated work. Economist Diane Coyle advocates expanded GDP measures that incorporate the value of digital services. Still, even proposed modifications ignore attention costs, the economic value destroyed by cognitive fragmentation, reduced well-being, and depleted mental health. A comprehensive attention accounting framework would require treating consciousness as a capital stock subject to depreciation through overuse and requiring investment for restoration, fundamentally challenging growth models assuming limitless intensification of mental labor.

The healthcare system provides particularly stark evidence of unaccounted attention costs. Physician burnout rates exceeding 50% correlate closely with the introduction of electronic health records, which mandate continuous data entry during patient encounters. Studies by economist Arnold Milstein analyzing physician time allocation reveal that doctors now spend two hours on documentation for every hour of patient contact, with much of this occurring after nominal work hours. This attention burden appears nowhere in hospital budgets or insurance reimbursements, instead manifesting as externalized costs through physician mental health deterioration, medical errors from cognitive overload, and early career exits, removing trained practitioners from the healthcare system. The economic irrationality becomes apparent only when adopting a full-cost accounting perspective, incorporating attention as a limited, depreciable asset requiring careful stewardship. Similar dynamics operate across knowledge work sectors where professional attention standards demand responsiveness incompatible with human cognitive limitations, creating systematic value destruction masked by incomplete economic measures.

The environmental economics literature offers instructive parallels for conceptualizing attention as a commons subject to tragedy. Ecologist Garrett Hardin's classic formulation identified how shared resources are prone to overexploitation when individual actors capture benefits from use while all commons users bear costs. Individual grazing animals on common pasture generate profit for their owners while degrading the collective resource incrementally. The attention commons operates identically: each advertisement, notification, and engagement hook extracts small amounts of individual attention while degrading the collective attentional environment through escalating baseline stimulation. Just as environmental regulation requires internalizing pollution costs through taxation or cap-and-trade systems, attention economics requires mechanisms forcing attention extractors to bear the full social costs their activities impose. Economist Ronald Coase's framework for addressing externalities through property rights assignment suggests potential regulatory approaches: individuals could possess legally protected attention rights that platforms must purchase rather than expropriating without compensation. Such frameworks remain largely theoretical, but growing recognition of attention as a finite resource with real economic value makes regulatory intervention increasingly plausible.

## Behavioral Economics and Attention Market Failures

Standard economic theory assumes rational actors making optimal choices given available information and preferences.

Behavioral economics demonstrates systematic deviations from rationality through cognitive biases and decision-making heuristics that attention markets exploit systematically. Hyperbolic discounting, the tendency to heavily weight immediate rewards over future benefits, makes present-moment attention capture extremely effective despite long-term costs. Watching one more video or scrolling through another screen of content provides immediate gratification so minuscule as to be ignored. At the same time, the accumulated opportunity cost of thousands of lost hours becomes apparent only retrospectively. Economist George Loewenstein's research on the hot-cold empathy gap shows that people systematically underestimate how future circumstances will affect their behavior. When not actively experiencing the pull of attention-capturing technologies, individuals genuinely believe they will resist these temptations, leading to systematic underinvestment in protective commitment devices such as app blockers or scheduled disconnection. Companies design choice architectures that exploit these biases: default settings maximize data sharing and engagement, cancellation processes involve multiple confirmation steps, creating friction absent from sign-up flows, and free trial periods leverage status quo bias, ensuring most users never actively cancel services.

The economic concept of "information asymmetry", where one party to a transaction possesses superior information that can be exploited, reaches its zenith in attention markets. Platforms employ teams of behavioral scientists, data analysts, and UX designers, using real-time experimental data to determine what effectively captures attention. Users lack both comparable expertise and visibility into the sophisticated persuasion techniques deployed against them. This creates principal-agent problems where platforms ostensibly serving user interests actually optimize for engagement metrics that benefit shareholders while harming users. Unlike traditional markets, where repeated interactions allow learning and reputation mechanisms to reduce asymmetry, attention platform dynamics resist these corrections. The constant novelty of content and interaction patterns prevents users from developing reliable heuristics, while algorithmic personalization means each user experiences a unique system, making collective knowledge-sharing difficult. Network effects and switching costs trap users even after recognizing exploitation, because the value of platforms depends on where other users direct their attention. An individual cannot unilaterally escape Facebook without sacrificing social connectivity, making rational individual action insufficient for market correction.

Economist Herbert Simon introduced the concept of "satisficing", seeking satisfactory rather than optimal solutions when search costs exceed potential gains. Attention markets systematically prevent satisficing through infinite scroll, autoplay, and algorithmic recommendation systems that eliminate natural stopping points. Traditional media contained built-in attention boundaries: television programs ended, requiring active channel selection; newspapers had final pages; and albums contained fixed track counts. These boundaries allowed satisficing; viewers watched a program, then moved to other activities, and readers finished the news section, then closed the paper. Digital platforms explicitly eliminate these stopping cues, continuously serving new content personalized to individual preferences, making the optimal stopping point perpetually uncertain. This exploits what economist Richard Thaler terms "the endowment effect", the tendency for people to overvalue things they possess or have invested time in. The sunk cost of attention already devoted to a platform makes disengagement psychologically costly even when continued use provides diminishing returns, creating an attention trap that rational decision-making would avoid but cognitive biases facilitate.

## Attention as Positional Good and Zero-Sum Competition

Economist Fred Hirsch introduced the concept of "positional goods" as products whose value derives primarily from relative scarcity rather than absolute utility. A Harvard degree provides value partly through exclusive prestige that would evaporate if universally available. Human attention functions as the ultimate positional good; individuals compete for attention from limited pools of potential awareness, making gains fundamentally zero-sum at the aggregate level. The creator economy exemplifies this: each successful influencer captures attention that could have gone to competitors, with no net increase in the system's total attention. This creates attention-capture arms races, where participants must continually escalate investment to maintain relative position, even in the absence of absolute gains. Political campaigns demonstrate this dynamic; candidates must match rivals' advertising expenditures not to gain attention but merely to avoid losing attention to competitors. Aggregate spending produces no net increase in voter attention while consuming resources that could generate actual social value, resulting in deadweight loss from positional competition. Economist Robert Frank documents how positional competition creates "expenditure cascades," in which top-tier competition drives spending throughout the distribution. The most prominent academic researchers receive disproportionate attention through citation advantages and conference invitations, pressuring others to prioritize self-promotion and visibility over research quality.

Social media transforms traditionally non-positional domains into attention competitions with status implications. Personal life updates once shared among friends for purely communicative purposes now get evaluated through comparative attention metrics, likes, shares, and comments, creating implicit ranking systems. Vacation photos become competitions for most impressive destinations; parenting updates compete for validation; even grief expressions get quantified through sympathy responses. This recasts everyday experience as a perpetual tournament requiring constant strategic attention management. Developmental psychologist Jean Twenge's analysis of adolescent mental health trends identifies sharp increases in anxiety and depression coinciding with smartphone adoption, plausibly caused by transforming peer relationships into quantified attention competitions with clear winners and losers. The economic insight reveals why individual-level solutions prove insufficient: even users who recognize the psychological costs cannot unilaterally withdraw without accepting lower social status and reduced access to relationships within attention-based social hierarchies.

The corporate world increasingly operates through attention-based status systems parallel to formal hierarchies. Email volume, meeting invitations, and Slack message rates signal organizational importance independent of actual authority or contribution. Employees strategically manage their attention visibility, responding promptly to high-status individuals, participating visibly in meetings, and maintaining active communication streams as survival strategies in attention-mediated corporate politics. This consumes substantial cognitive resources that could contribute to productive work, representing another form of positional competition waste. Economist Thorstein Veblen identified similar dynamics a century ago through the concept of "conspicuous consumption", purchasing luxury goods primarily to signal status rather than derive intrinsic utility. Contemporary professionals engage in conspicuous availability, perpetual responsiveness, and attention accessibility, signaling dedication and indispensability. Just as Veblen's leisure class competed through ostentatious waste of material resources, the contemporary professional class competes through ostentatious waste of attention and time, with predictable results for wellbeing and productivity.

The market for attention has fundamental characteristics that produce inefficient equilibria that individual rationality cannot overcome. Game theory identifies this as a coordination problem: all players would benefit from mutual attention conservation, but without binding agreements, the individually rational strategy involves defection through continued attention competition. The solution requires collective action through social norms, institutional rules, or regulatory intervention that shifts the equilibrium toward mutually beneficial attention stewardship. Some organizations experiment with such interventions: companies implementing "no meeting days" preserve focused work time; email curfews reduce after-hours attention demands; asynchronous communication norms reduce real-time availability requirements. These policies work only when adopted uniformly, preventing defection by individuals or teams that gain a competitive advantage by violating attention boundaries. The economic challenge involves making attention conservation the rational individual strategy rather than requiring altruistic self-sacrifice, demanding changes to the incentive structures that currently reward attention extraction regardless of social costs.

The economics of attention reveals that contemporary attention dysfunction stems not primarily from individual weakness or poor personal choices but from rational responses to market structures that systematically misalign individual and collective interests. When platforms profit from attention capture regardless of user wellbeing, when workers must demonstrate constant availability to maintain employment security, when social status depends on attention competition success, individuals acting rationally within these systems produce collectively destructive outcomes. Understanding attention through economic frameworks illuminates why personal productivity advice and mindfulness practices, while potentially valuable for individuals, cannot address systemic problems requiring institutional and regulatory solutions. The attention economy represents a market failure that demands intervention, through strengthened consumer protection, platform regulation, workplace standards, or new economic measures that incorporate attention costs, if human cognitive flourishing rather than mere engagement metrics is to guide technological development and social organization.

# Chapter 10: Art and Attention: The Power of Visual Storytelling

When Pablo Picasso unveiled "Guernica" in 1937, viewers did not merely glance at the canvas and move on. The painting demanded something far more profound; it commandeered attention through calculated visual disruption, forcing observers to linger, to struggle with its fragmented forms, to construct meaning from deliberate chaos. The mural's monochromatic palette eliminated color as a distraction, channeling all perceptual resources toward compositional relationships. Its layered imagery, the screaming horse, the broken warrior, the mother clutching her dead child, created what art historians term "visual density," where each region contains multiple competing narratives that prevent casual scanning. This density generates a peculiar temporal experience: the artwork cannot be consumed in a single viewing but requires sustained, recursive examination, in which each return reveals previously overlooked elements. Picasso understood intuitively what cognitive science would later confirm, that visual art possesses unique capacities to capture, manipulate, and sustain attention through deliberate exploitation of perceptual processing mechanisms. Unlike linguistic communication, which unfolds linearly over time, or musical expression, which exists purely in temporal sequence, visual art presents simultaneous information across spatial dimensions, creating attentional demands fundamentally different from those of other expressive modalities. The painter, photographer, or filmmaker commands not just what viewers see but how long they look, where their gaze travels, what emotional states accompany perception, and ultimately what cognitive residue remains after the encounter concludes.

The relationship between art and attention transcends aesthetic appreciation, extending into domains of persuasion, memory formation, and cultural transmission, where visual storytelling determines what information survives in collective consciousness. Documentary photographer Sebastião Salgado spent eight years creating "Genesis," a photographic project depicting pristine ecosystems and indigenous communities. The resulting images employ compositional techniques specifically designed to overcome the attention-numbing effect of environmental catastrophe fatigue. Rather than showing overt destruction, melting glaciers, clear-cut forests, and polluted rivers, Salgado photographed what remains intact, creating visual experiences of such arresting beauty that viewers spontaneously generate protective impulses. This strategic reversal exploits what environmental psychologist Stephan Kellert termed "biophilia", the innate human affinity for natural forms and living systems that captures attention without conscious effort. The photographs function as attentional Trojan horses: viewers believe they are simply appreciating beautiful landscapes, while unconsciously encoding powerful emotional associations linking natural preservation with aesthetic reward. Subsequent studies measuring donation behavior and environmental policy support found that exposure to Salgado's positive imagery generated significantly more decisive pro-environmental action than conventional documentation of ecological damage. The mechanism operates through what neuroscientist Antonio Damasio describes as "somatic markers", embodied emotional responses that guide decision-making beneath conscious awareness. Beautiful nature imagery creates positive somatic markers that predispose viewers toward conservation choices when later facing related decisions, demonstrating how visual art influences behavior through attention capture that bypasses rational deliberation entirely.

## The Architecture of Visual Attention: How Images Guide the Eye

Eye-tracking technology reveals that visual attention follows predictable pathways determined by image composition, chromatic relationships, and semantic content hierarchies. When viewers encounter a portrait photograph, fixation patterns typically follow the "face-bias" sequence: eyes first, then mouth, then overall facial structure, before potentially moving to background elements. This hardwired preference emerges from evolutionary selection favoring rapid social threat assessment, determining whether encountered faces signal danger or safety, which has determined survival outcomes for hundreds of millennia. Contemporary visual artists exploit this biological inheritance through compositional choices that either honor or deliberately violate expected scanning patterns. Dutch Golden Age painters like Johannes Vermeer positioned human faces at compositional power points, locations determined by golden ratio proportions where the eye naturally gravitates, ensuring viewers' attention landed precisely where narrative significance concentrated. Conversely, modernist photographers like Lee Friedlander deliberately placed human subjects at compositional peripheries while foregrounding mundane objects, creating cognitive dissonance where automatic face-seeking behavior encountered unexpected visual hierarchies. This violation generates sustained attention as perceptual systems attempt to resolve the contradiction between biological priorities and compositional logic, effectively trapping consciousness in extended processing that far exceeds typical viewing duration.

The temporal unfolding of visual attention depends critically on image complexity and semantic coherence. Cognitive scientist Stephen Palmer demonstrated, through reaction-time experiments, that coherent scenes, images in which all elements relate through consistent spatial logic and thematic unity, permit rapid gist extraction within 200 milliseconds, allowing viewers to categorize content almost instantly. However, paintings and photographs designed for aesthetic impact systematically violate coherence principles to prevent premature categorization that would terminate attentive viewing. Surrealist photographer Philippe Halsman's "Dalí Atomicus" depicts Salvador Dalí suspended mid-leap surrounded by floating cats, water, and easels, a scene impossible in physical reality yet rendered with photographic verisimilitude. The image creates categorical ambiguity, hindering rapid gist extraction: Is this documentation or fantasy? Sculpture or photograph? Humor or profundity? This semantic indeterminacy sustains attention by preventing the closure that normally terminates visual processing once categorization succeeds. Art historian Ernst Gombrich termed this principle "the beholder's share", the necessity for viewers to actively construct meaning rather than passively receiving it. Images that provide all meaning immediately become aesthetically inert, while those demanding interpretive labor generate the sustained engagement that distinguishes art from mere visual information.

## Color Theory and Attentional Arousal States

Chromatic choices in visual art directly modulate arousal via neurophysiological pathways linking retinal processing to limbic structures that govern emotional states and attentional readiness. The human visual system processes color through opponent channels, red-green, blue-yellow, and light-dark, with each pairing creating perceptual tension when complementary hues appear in proximity. When Mark Rothko painted "Orange and Yellow," he positioned these adjacent colors at an enormous scale, creating chromatic vibrations in which the boundaries between hues appear to pulsate or dissolve. This phenomenon, termed "chromostereopsis," results from the differential focal lengths required to perceive different wavelengths: long wavelengths, such as red and orange, focus behind the retina, while short wavelengths, such as blue, focus in front, creating perceived depth relationships that shift continuously as the eye adjusts. The resulting perceptual instability prevents the visual system from settling into a steady state, maintaining heightened attentional engagement through bottom-up capture mechanisms that register continuous change. Rothko understood that chromatic activation could substitute for representational content in commanding attention, using color relationships alone to generate viewing experiences as temporally extended and emotionally charged as figurative narrative painting.

Cultural context profoundly influences chromatic attention capture through learned color-meaning associations that vary dramatically across societies. In Western contexts, red activates attentional systems through associations with danger, arousal, and urgency, drawn from both biological threat detection and cultural conventions such as stop signs and warning labels. Experimental research by psychologist Andrew Elliot found that merely presenting achievement tasks on red backgrounds impaired performance compared to neutral colors, suggesting that color-arousal associations operate automatically outside conscious control. Yet in traditional Chinese culture, red signifies prosperity, celebration, and good fortune, creating opposite attentional and emotional responses. Contemporary photographer Steve McCurry exploits these cross-cultural chromatic vocabularies in his documentary work, deliberately seeking moments in which color elicits universal rather than culturally specific responses. His famous "Afghan Girl" portrait employs piercing green eyes against red fabric. This combination captures Western attention through complementary color contrast while simultaneously signaling cultural specificity through the girl's traditional dress. This chromatic bilingualism allows single images to operate simultaneously across multiple cultural attention systems, demonstrating how sophisticated visual artists manipulate not just perception but the complex interplay between hardwired perceptual mechanisms and culturally constructed interpretive frameworks.

The saturation and luminance dimensions of color independently influence attentional capture beyond hue alone. Highly saturated colors, pure, intense chromatic values unmixed with gray, trigger stronger orienting responses than desaturated tones, explaining why advertisement design invariably employs vivid colors competing for limited attentional resources in cluttered visual environments. However, artistic practice often deliberately employs desaturation to modulate attention away from immediate capture toward sustained contemplation. Ansel Adams's black-and-white landscape photographs eliminate chromatic information, forcing viewers to engage tonal relationships, compositional geometry, and textural details that color's attentional dominance might otherwise obscure. This strategic subtraction reveals an essential principle of visual attention management: sometimes capturing attention requires not adding stimulation but removing competing signals that prevent deeper perceptual engagement. The monochromatic palette creates what photographer John Szarkowski termed "visual silence", a reduction of sensory noise that paradoxically enhances rather than diminishes perceptual intensity by focusing limited attentional resources on fewer information channels.

## Narrative Sequence and Cinematic Attention Control

Motion picture editing demonstrates the most sophisticated attention manipulation techniques ever developed, representing over a century of systematic experimentation in consciousness steering through visual sequence construction. When Soviet filmmaker Lev Kuleshov intercut identical shots of actor Ivan Mosjoukhin's neutral expression with images of soup, a dead woman in a coffin, and a child at play, audiences reported seeing distinct emotions, hunger, grief, and tenderness in the unchanging face. This "Kuleshov Effect" revealed that meaning and emotional response emerge not from individual images but from juxtaposition relationships, with viewers' attention actively constructing narrative coherence by inferring causality between sequential shots. Contemporary film editors exploit this principle through "emotional geography", manipulating shot duration, composition, and sequence to guide viewers through precisely orchestrated attentional and affective states. Action sequences employ rapid cutting with average shot lengths under two seconds, overwhelming conscious processing to generate visceral arousal. Contemplative dramas use extended shots lasting thirty seconds or more, allowing viewers' attention to settle into sustained observation that mirrors meditative awareness. The editor controls not merely what viewers see but their temporal experience of consciousness itself, accelerating or decelerating subjective time through cutting rhythm.

The "180-degree rule" in cinematography demonstrates how spatial continuity maintains viewer orientation and prevents disruptive attentional resets. This convention establishes an imaginary axis between characters in a scene, with the camera restricted to one side of it across shot changes.

Maintaining consistent screen direction, if Character A faces screen-left toward Character B in one shot, the reverse angle must show Character B facing screen-right toward Character A, allowing viewers' attention to track the conversation without reconstructing spatial relationships after each cut. Violating this rule, termed "crossing the line," disorients viewers as characters appear to switch positions, forcing attentional resources toward resolving spatial confusion rather than processing narrative content. Yet some filmmakers deliberately cross the line precisely for this disorienting effect. Stanley Kubrick in "The Shining" systematically violated continuity conventions to create subliminal unease, with viewers' perceptual systems detecting impossible architectural relationships and spatial inconsistencies that generated persistent low-level anxiety without conscious awareness of the source. This demonstrates how mastery of attention mechanics enables filmmakers to manipulate not just what viewers consciously perceive, but also the background emotional tone that colors entire viewing experiences.

The shot-reverse-shot pattern governing filmed conversations reveals how editing simulates natural attention shifts during face-to-face interaction. When watching two people converse, our gaze naturally alternates between speakers following conversational turn-taking. Film editing replicates this through cutting between characters as they speak, allowing viewers' attention to follow the flow of dialogue without the effort of voluntarily redirecting the gaze. However, this convention's ubiquity renders it invisible; viewers experience shot-reverse-shot sequences as transparent windows onto events rather than as constructed attentional guidance. Experimental filmmaker Chantal Akerman rejected this transparency in "Jeanne Dielman," employing static wide shots in which characters enter, perform actions, and exit the frame without cutting. The resulting viewing experience proves initially frustrating as viewers must voluntarily direct attention within fixed frames rather than following editorial guidance. Yet this frustration transforms into heightened awareness, and viewers become conscious of their own attention as an active process rather than a passive reception. Akerman's formal strategy thus converts cinema from an attention-capture device into an attention-training tool, forcing viewers to develop voluntary control over perceptual focus comparable to that in meditation practice. The radical slowness and editorial refusal make visible the usually invisible attentional labor that all viewing entails, revealing how conventional cinema's seeming effortlessness actually represents sophisticated manipulation rather than natural perception.

## Photography's Decisive Moment: Capturing Transient Attention

Henri Cartier-Bresson's concept of "the decisive moment" articulates photography's unique relationship to temporal attention, the capacity to isolate singular instants from experience's continuous flow and render them permanently available for extended contemplation. Cartier-Bresson's "Behind the Gare Saint-Lazare" freezes a man mid-leap over a puddle, his reflected shadow visible in the water below. The photograph captures an instant lasting perhaps one-tenth of a second. This moment would register only subliminally in unmediated perception, as our attention cannot fixate on such brief events. Yet photography's temporal arrest allows unlimited viewing duration, enabling attention to thoroughly explore compositional relationships, the geometric echo between leaping figure and poster fragments, the tonal balance between light and shadow, and the tension between motion frozen and water's stillness. This temporal paradox generates what Roland Barthes termed "punctum", details that emotionally pierce viewers despite having no apparent narrative significance. In Cartier-Bresson's image, the punctum might be the blurred figure in the background, the torn poster's cryptic text, or the puddle's irregular shape, elements that capture individual attention through mysterious personal resonance rather than formal composition or documentary content. Photography thus creates attentional experiences impossible in direct perception: extended meditation on ephemeral moments, emotional engagement with strangers never encountered, and aesthetic appreciation of quotidian scenes ordinary attention dismisses as unworthy of notice.

Contemporary photographer Rinko Kawauchi employs what she terms "small measures", extremely subtle subject matter, photographed with technical precision, that forces viewers' attention to overlooked aspects of daily experience. Her images depict soap bubbles, light through curtains, and children's hands holding insects, scenes that require no technical manipulation yet generate profound attentional shifts through simple acts of noticing and framing. By photographing the perpetually ignored, Kawauchi performs what phenomenologist Maurice Merleau-Ponty described as "returning to the things themselves", breaking through habituated perception that renders familiar environments invisible through overfamiliarity. Her work demonstrates that visual art's attention power derives not necessarily from spectacle or technical virtuosity but from artists' capacity to perceive freshly and thereby train viewers' attention toward renewed perceptual awareness. This pedagogical function of visual art, teaching us how to see rather than what to see, may represent its most profound contribution to attention cultivation, offering systematic training in breaking the attentional autopilot that governs typical perception.

The seriality principle in contemporary photography creates attentional experiences unachievable through single images. Bernd and Hilla Becher photographed industrial structures, water towers, cooling towers, grain elevators, over decades, presenting them in gridded typological arrangements in which subtle variations become visible through repetition. A single water tower photograph might register as a documentary record or a formalist composition, but viewing sixty similar structures arranged systematically transforms attention from individual forms to underlying patterns, variations within type, and design evolution across time and location. This serial approach trains attention toward discrimination rather than recognition, developing the capacity to perceive subtle differences within apparent similarity rather than crude categorical identification. Cognitive psychologist Eleanor Rosch demonstrated that expertise in any domain involves progressive differentiation of initially monolithic categories: novice bird watchers see "small brown birds," while experts immediately distinguish between sparrow species through plumage details, flight patterns, and habitat preferences. Photographic seriality accelerates this differentiation process, systematically exposing viewers to variations that build perceptual schemas supporting nuanced attention to previously undifferentiated phenomena.

## Visual Metaphor and Symbolic Attention Capture

The surrealist movement pioneered the construction of visual metaphors that capture attention through semantic impossibility, images depicting scenes that cannot exist physically yet appear photographically real. René Magritte's "The Son of Man" presents a man in a bowler hat whose face is obscured by a floating green apple. The painting generates sustained attention through categorical violation: apples do not levitate, and if they did, why would they position themselves precisely to obscure faces? This impossibility prevents perceptual closure, trapping attention in unsuccessful attempts at coherent interpretation. Yet the semantic instability proves generative rather than merely frustrating, viewers construct metaphorical meanings connecting concealment, identity, and the unknowability of others. The apple becomes simultaneously a barrier and an invitation, obscuring the man's identity while paradoxically making viewers hyperaware of his hidden interiority. Magritte understood that visual metaphor operates through controlled ambiguity, providing sufficient structure for interpretation while withholding the determinacy that would terminate attentive viewing. This principle extends beyond surrealism into advertising, political propaganda, and social movement iconography, wherever visual communication aims not just at information transmission but at sustained attentional engagement that allows messages to penetrate beyond conscious processing into memory formation and belief modification.

Symbolic visual communication demonstrates attention capture through the activation of cultural codes that operate automatically in acculturated viewers. The Christian cross, Islamic crescent, or corporate logos like Nike's swoosh trigger immediate recognition responses requiring minimal processing resources, what cognitive scientist Lawrence Barsalou terms "perceptual symbols" that activate entire conceptual networks through single visual forms. Artist Barbara Kruger exploits this symbolic efficiency in her text-image works, superimposing declarative statements over appropriated photographs. Her piece "Your Body Is a Battleground" combines this text with a woman's bisected face, one half positive, the other negative. The symbolic density, body, battlefield, division, and gender activate multiple interpretive frameworks simultaneously, generating attentional engagement through semantic richness that rewards extended contemplation. Unlike didactic messaging that exhausts meaning through single viewing, Kruger's work maintains interpretive productivity across repeated encounters as viewers discover new conceptual connections between image and text components. This inexhaustibility represents visual art's most valuable attentional property: the capacity to sustain engagement not through novelty but through structured complexity that supports perpetual rediscovery.

The relationship between visual art and attention ultimately reveals that seeing constitutes not passive reception but active construction, with artists providing structured occasions for conscious acts of attention that might otherwise never occur. When Georgia O'Keeffe painted flowers at monumental scale, she explicitly aimed to force urbanites' attention toward forms they routinely overlooked. As she wrote, "Nobody sees a flower, really, it is so small, we haven't time, and to see takes time, like to have a friend takes time." Her enlarged irises and jimson weeds command attention through sheer scale. Still, more profoundly, they teach that genuine perception requires time allocation: truly seeing something means granting it sustained attentional resources typically reserved for the most instrumentally valuable objects. This pedagogical dimension, teaching viewers that attention itself constitutes a gift, a form of care, and a prerequisite for genuine understanding, may represent visual art's deepest function in an economy that systematically extracts and commodifies consciousness. Art offers not escape from attention demands but a sacred space where attention becomes voluntary rather than captured, where we choose what consciousness illuminates rather than surrendering that choice to algorithmic optimization. The painter's canvas, the photographer's frame, the filmmaker's sequence, these serve as training grounds where we practice the very capacity most threatened by contemporary attention economics: the ability to direct our own awareness according to our deepest values rather than others' commercial imperatives.

# Chapter 11: Lost in Thought: The Benefits of Daydreaming

The mathematician Andrew Wiles spent seven years working on Fermat's Last Theorem in complete secrecy, filling notebooks with failed approaches and abandoned proofs. Yet the breakthrough that finally unlocked the solution arrived not during intense calculation but while he sat in his office, staring absently at nothing in particular, his mind drifting through mathematical landscapes without conscious direction. This moment of productive inattention represents a profound paradox in our understanding of consciousness: sometimes the most valuable cognitive work occurs precisely when we stop trying to think. The phenomenon challenges every cultural narrative that equates mental productivity with focused concentration, revealing that the human brain requires periods of apparent idleness to perform its most sophisticated operations. Contemporary neuroscience has begun mapping what happens during these episodes of mental wandering, discovering not cognitive laziness but rather complex neural activity that consolidates memories, generates insights, simulates future scenarios, and maintains psychological coherence. The wandering mind, far from representing attention failure, constitutes an essential mode of consciousness that evolution has preserved precisely because it provides cognitive benefits unavailable through directed thought alone.

The discovery that minds naturally drift away from present tasks emerged from some of psychology's most tedious experiments. In the 1960s, researcher Jerome Singer asked participants to press buttons in response to random tone sequences while simultaneously reporting whenever they noticed their thoughts had strayed from the task. The results proved startling: even during this simple activity requiring minimal cognitive load, participants' minds wandered approximately thirty to forty percent of the time. Subsequent research employing experience sampling methods, randomly prompting people throughout their daily lives to report their current mental contents, revealed even higher rates. Psychologists Matthew Killingsworth and Daniel Gilbert created a smartphone application that contacted users at unpredictable intervals, asking what they were doing, what they were thinking about, and how they felt. Analyzing over 250,000 responses from thousands of participants, they found that minds wander forty-seven percent of waking hours. This was not an occasional distraction but the brain's default operating mode, suggesting that undirected thought serves fundamental purposes rather than representing mere attention failure. The mind, it seems, possesses an intrinsic restlessness that continually pulls awareness away from immediate perception toward self-generated mental content. This tendency persists across cultures, age groups, and activity types, pointing toward deep evolutionary origins rather than modern pathology.

## The Autobiographical Function: Constructing Coherent Selves

One crucial function of mental wandering is the consolidation of autobiographical memory and the construction of self-narratives. During periods when attention detaches from external demands, the brain engages in what psychologist Dan McAdams calls "life story processing", reviewing past experiences, identifying patterns across seemingly disconnected events, and weaving individual episodes into coherent narratives that define personal identity. This process cannot occur during focused external attention because it requires simultaneously holding multiple memories in consciousness while comparing them, extracting common themes, and evaluating their significance for understanding who one is. Research by neuroscientist Kalina Christoff using real-time fMRI during mind-wandering episodes revealed intense activity in the medial prefrontal cortex and posterior cingulate cortex, regions associated with self-referential processing and autobiographical recall. Critically, this activity showed stronger connectivity between brain regions during mind-wandering than during rest or focused tasks, suggesting active integration rather than passive disengagement.

The construction of temporal identity, maintaining a continuous sense of self despite constant change, depends fundamentally on these wandering episodes. Without regular opportunities to mentally review the past and project into the future, experience fragments into disconnected moments lacking narrative coherence. Patients with severe attention deficits who cannot sustain mind-wandering report feeling psychologically unmoored, experiencing life as a series of present moments without meaningful connection to personal history or future aspirations. Philosopher Galen Strawson has argued that some individuals lack this narrative self-structure entirely, experiencing themselves as momentary subjects without extended temporal identity. However, whether this represents a cognitive difference or a deficit in mind-wandering capacity remains contested. For most people, however, the autobiographical self emerges precisely through these periods of apparent inattention where consciousness turns inward to examine its own trajectory across time.

Cultural practices surrounding solitude implicitly recognize what neuroscience now demonstrates explicitly: periods of undirected thought are essential for maintaining psychological integration. The religious practice of retreat, temporarily withdrawing from ordinary activities into environments minimizing external stimulation, creates protected time for extensive mind-wandering. Participants consistently report profound experiences of self-understanding, shifts in perspective on ongoing life situations, and emotional resolution of past difficulties. These outcomes emerge not through deliberate analytical thinking but through allowing consciousness to freely associate across memories, feelings, and hypothetical scenarios without goal-directed constraints. The writer Virginia Woolf, who struggled throughout her life with what she termed "the cotton wool of daily life," created elaborate daily routines ensuring several hours of unstructured time for walking or simply sitting, recognizing that her creative work depended absolutely on these periods when her mind could roam without purpose. Modern productivity culture, with its relentless emphasis on optimizing every moment, systematically eliminates precisely these unstructured intervals that make sophisticated self-understanding possible.

## Future Simulation and Adaptive Planning

Beyond consolidating experience, mind-wandering enables prospective thinking, mentally simulating potential futures to evaluate possible actions and prepare for anticipated challenges. Cognitive scientist Daniel Schacter's research revealed that remembering the past and imagining the future activate remarkably similar neural networks, suggesting that both processes draw on standard mechanisms for constructing mental scenarios. This "prospective brain" function operates primarily during mind-wandering episodes, when consciousness generates spontaneous simulations: imagining upcoming conversations, rehearsing difficult discussions, anticipating obstacles to planned activities, or exploring counterfactual scenarios in which different choices lead to alternative outcomes. These mental simulations provide low-cost opportunities to practice behaviors, identify potential problems, and refine strategies before encountering actual situations. Professional athletes have long exploited this through visualization training, but neuroscience reveals that all humans engage in constant spontaneous future simulation during periods of mental drift.

The evolutionary logic becomes clear: organisms capable of mentally simulating multiple futures and evaluating potential consequences possess enormous adaptive advantages over those limited to reactive response. An early human whose mind wandered to imagining tomorrow's hunt, mentally rehearsing stalk patterns, anticipating prey behavior, and considering what could go wrong, gained preparation that improved actual hunting success. This capacity for mental time travel, as psychologist Thomas Suddendorf terms it, represents one of humanity's most distinctive cognitive achievements. Critically, these simulations cannot be commanded deliberately with the same richness they achieve during spontaneous mind-wandering. When consciously planning, we generate limited, logical scenarios constrained by current knowledge and explicit reasoning. Mind-wandering, conversely, produces far more elaborate simulations incorporating emotional dimensions, social dynamics, and unexpected possibilities that conscious planning excludes. The wandering mind essentially runs predictive models of reality, generating insights about future opportunities that become available to conscious reflection once attention returns to directed thought.

Studies examining people's spontaneous thoughts throughout the day reveal that approximately thirty percent of mind-wandering content involves future events, with the temporal distance ranging from hours ahead to decades into the future. Interestingly, these projections show systematic biases: people overestimate the likelihood of positive outcomes while underestimating obstacles, a pattern psychologist Tali Sharot calls the "optimism bias." While this might seem irrational, it serves crucial motivational functions; overly realistic simulations might paralyze action through excessive attention to potential failures. The slight positive distortion in prospective thinking during mind-wandering appears calibrated to balance realism with the motivational confidence needed to pursue goals.

Additionally, research by psychologist Gabriele Oettingen demonstrates that people who engage in elaborate positive future fantasizing during mind-wandering paradoxically achieve worse outcomes than those employing "mental contrasting", spontaneously alternating between optimistic scenarios and realistic obstacle consideration. This suggests effective mind-wandering involves not pure fantasy but iterative simulation that explores multiple possibilities, including difficulties, thereby preparing consciousness for adaptive response across varied circumstances.

## Creative Insight and Remote Association

Perhaps the most celebrated benefit of mind-wandering involves creative insight, the sudden perception of connections between previously unrelated concepts that produce novel understanding or solutions. The phenomenon appears throughout creative domains: scientific discoveries attributed to dreams or reverie, artistic inspiration emerging during walks or baths, and business innovations conceived while their creators were supposedly relaxing. While these anecdotes might seem coincidental, laboratory research confirms that mind-wandering directly enhances creative thinking through mechanisms unavailable during focused concentration. Psychologist Jonathan Schooler's experiments demonstrated that participants who performed tasks designed to induce mind-wandering subsequently showed improved performance on creativity tests requiring unusual word associations and novel uses of everyday objects. The effect proved specific to tasks requiring conceptual combination rather than analytical reasoning, suggesting mind-wandering facilitates a particular cognitive mode.

The mechanism involves what creativity researcher Mark Jung-Beeman calls "semantic spreading activation"; during mind-wandering, mental associations propagate more broadly across memory networks than during focused thought. Usually, directed attention constrains associative processing to concepts closely related to the current task, preventing irrelevant information from interfering with goal pursuit. This proves advantageous for execution but limits the discovery of remote connections that enable insight. Mind-wandering releases these constraints, allowing activation to spread through unexpected pathways that link distantly related concepts. Neurologically, this corresponds to reduced activity in cognitive control regions that typically suppress tangential associations, combined with heightened activity in associative cortices that store conceptual knowledge. The result resembles exploratory search through semantic networks, where unusual combinations spontaneously emerge into consciousness. When one of these combinations proves relevant to a problem the conscious mind has been contemplating, the experience manifests as sudden insight, the famous "Aha!" moment that feels spontaneous precisely because it emerged from unconscious processing during apparent inattention.

Composer John Cage designed musical pieces that explicitly incorporated extended silences to create space for listeners' minds to wander and generate their own mental content, recognizing that reception involves not just absorbing transmitted information but also combining it with spontaneous mental associations. Similarly, architect Christopher Alexander argued that significant buildings succeed not through forcing attention along predetermined paths but through creating spaces that invite reverie, window seats overlooking gardens, corridors with surprising views, and rooms whose proportions encourage contemplative lingering. These design principles recognize that psychological value often emerges not from controlling attention but from liberating it to make unexpected connections. The modern workplace, with its emphasis on eliminating "wasted time" through continuous meetings and measurement, systematically destroys the conditions necessary for this associative creativity. Software developer Jason Fried observed that his most productive programmers accomplished their best work during undisturbed periods lasting multiple hours, not because they maintained perfect focus throughout, but because extended uninterrupted time allowed natural cycles of concentration and mind-wandering that enabled both execution and insight.

## Emotional Regulation and Psychological Distance

Mind-wandering serves crucial emotional regulation functions by allowing consciousness to revisit distressing experiences with sufficient psychological distance to reprocess them without overwhelming activation. Psychologist Ethan Kross discovered that people spontaneously adopt a distanced perspective during mind-wandering about emotional events, mentally viewing situations from a third-person perspective or using their own name instead of "I" when thinking about themselves. This spontaneous distancing reduces emotional intensity while preserving the content of memory, facilitating what therapists call "emotional digestion." Without these periods of spontaneous processing, distressing experiences remain cognitively unintegrated, intruding repeatedly into consciousness with full emotional force rather than gradually losing their acute sting. The phenomenon explains why insomnia, which prevents the mind-wandering that occurs during the transitional states between sleeping and waking, exacerbates emotional disorders. Individuals forced to maintain constant external attention never achieve the psychological distance necessary for emotional experiences to resolve.

The process operates particularly during what sleep researchers call "quiet wakefulness", periods when people remain conscious but not engaged in demanding tasks, allowing their minds to wander freely over recent experiences. Research using fMRI during post-task rest periods revealed that the brain spontaneously replays neural activity patterns from earlier emotional experiences, with reduced amygdala activation compared to the original events. This suggests the brain uses mind-wandering episodes to repeatedly reactivate emotional memories in safe contexts, gradually weakening their capacity to trigger intense reactions. Trauma therapy techniques like Eye Movement Desensitization and Reprocessing (EMDR) may work partially by facilitating this natural process through structured protocols that encourage distanced mental processing. However, the spontaneous version occurring during ordinary mind-wandering performs similar functions without clinical intervention, provided individuals receive sufficient opportunity for undirected thought.

Cultural differences in tolerance for boredom and mental idleness may reflect an implicit understanding of the emotional regulation benefits they confer. Mediterranean cultures traditionally embrace extended periods of apparent inactivity, the Spanish siesta, the Italian passeggiata, which create protected time for mental wandering. Participants in these practices report not just physical rest but psychological restoration, emerging with renewed emotional equilibrium. Conversely, cultures that emphasize constant productivity and view idle time as a moral failure inadvertently eliminate opportunities for spontaneous emotional processing. Psychologist Sandi Mann's research found that people in such cultures show higher rates of anxiety and depression, potentially because relentless activity prevents the mind-wandering necessary for emotional regulation. The contemporary epidemic of burnout may reflect not just overwork but insufficient cognitive downtime; people remain busy every waking moment through either work demands or digital entertainment, never experiencing the seemingly vacant periods when emotional processing naturally occurs.

## The Incubation Effect and Problem-Solving

When confronting complex problems, deliberately ceasing active work and allowing the mind to wander often produces better solutions than sustained, focused effort. This "incubation effect" has puzzled researchers since Dutch psychologist Ap Dijksterhuis first systematically demonstrated it: participants asked to choose between complex options made better decisions after distraction periods than after equivalent time spent in deliberate analysis. The counterintuitive finding challenged the assumption that more thinking automatically produces better outcomes, suggesting that certain cognitive operations require periods of inattention to succeed. Subsequent research refined the finding: incubation benefits appear primarily for problems requiring insight or creative reframing rather than systematic analysis. Mathematical proofs, chess positions, or logical deductions benefit little from incubation. At the same time, problems requiring novel perspectives, designing innovative products, resolving interpersonal conflicts, or recognizing hidden assumptions show dramatic improvement after mind-wandering intervals.

The mechanism likely involves continued unconscious processing during mind-wandering that explores solution spaces more freely than conscious deliberation permits. Focused problem-solving tends toward incremental refinement of initially adopted approaches, rarely abandoning frameworks even when they prove inadequate. Mind-wandering, by contrast, allows mental representations to restructure without the constraints conscious attention imposes. Cognitive scientist Stellan Ohlsson describes insight as "representational change", suddenly perceiving a problem through a different conceptual lens that makes previously hidden solutions obvious. These changes occur more readily when attention has moved away, releasing cognitive fixation on unproductive approaches. The mathematician Henri Poincaré described exactly this pattern in his own work. After days of fruitless calculation, he would abandon the problem and go for a walk, during which the solution would spontaneously appear in consciousness fully formed. He attributed this to unconscious work continuing during the conscious mind's inattention, a hypothesis neuroscience now supports through evidence of ongoing problem-relevant neural activity during mind-wandering episodes following concentrated work on complex tasks.

Organizations increasingly recognize the benefits of incubation through policies that protect unstructured time. Technology companies like Google famously allowed engineers twenty percent time for personal projects, explicitly acknowledging that minds need freedom to wander toward unexpected connections. While such policies often face pressure during economic downturns, companies that maintain them report disproportionate innovation output, suggesting the monetary value of sanctioned mind-wandering exceeds the lost focused productivity. Similarly, academic research culture traditionally protected faculty time through tenure systems and light teaching loads, recognizing that intellectual breakthroughs require extended periods of apparent idleness. Contemporary pressures toward measurable productivity increasingly threaten these arrangements, potentially eliminating the cognitive slack necessary for transformative insights while optimizing for incremental outputs.

## Individual Differences and Mind-Wandering Patterns

Not all mental wandering is equally beneficial, and substantial individual differences exist in both the frequency and the content of spontaneous thought. Psychologist Jonathan Smallwood distinguishes between deliberate mind-wandering, intentionally allowing thoughts to drift, and spontaneous mind-wandering that occurs despite efforts to maintain focus. While both types correlate with creativity benefits, they show different relationships with well-being. Deliberate mind-wandering during appropriate contexts associates with positive mood and life satisfaction, whereas frequent spontaneous mind-wandering during tasks requiring attention predicts negative affect and decreased performance. This suggests that mind-wandering benefits depend critically on context appropriateness: mental drift proves valuable during activities that tolerate divided attention but problematic during safety-critical tasks like driving or high-stakes concentration work.

Content matters as much as frequency. Research distinguishes between constructive mind-wandering, thoughts about planning, problem-solving, or positive memories, and ruminative mind-wandering focused on adverse past events or anxious future scenarios. Clinical psychologist Sonja Lyubomirsky's extensive work on rumination reveals that while constructive mind-wandering enhances psychological well-being, rumination maintains depression and anxiety through repetitive dwelling on distress without generating valuable insights. The difference lies not in whether minds wander but in the valence and productivity of mental content. People who show a tendency toward rumination essentially have minds that repeatedly walk into the same hostile territories, preventing the breadth of exploration that makes mind-wandering valuable. Cognitive therapy techniques targeting rumination essentially teach patients to notice when mind-wandering has become unproductive and gently redirect attention toward more constructive mental content or external engagement.

Developmental changes in mind-wandering capacity illuminate its cognitive sophistication. Young children show limited capacity for sustained mind-wandering, with mental drift consisting primarily of immediate perceptual distraction rather than elaborate mental scenario generation. The ability to construct complex imaginary scenarios during mind-wandering develops gradually through childhood, paralleling growth in executive function and working memory capacity. This suggests mind-wandering requires substantial cognitive resources, sufficient working memory to maintain imagined scenarios, adequate executive function to disengage from perception and sustain self-generated thought, and developed autobiographical memory to provide material for mental simulation. Far from representing attention failure, effective mind-wandering demonstrates cognitive maturity. Elderly individuals experiencing cognitive decline show reduced mind-wandering frequency and more straightforward mental content, with progression tracking overall executive dysfunction. The loss diminishes quality of life beyond obvious cognitive impacts, as patients report feeling psychologically impoverished, unable to mentally time-travel, generate elaborate plans, or engage in the rich mental life that characterizes human consciousness.

## Cultivating Productive Mind-Wandering

Recognizing mind-wandering's benefits raises questions about deliberately cultivating it without tipping into maladaptive rumination or problematic inattention during demanding tasks. Several practices appear to optimize the balance. First, scheduling protected time explicitly for undirected thought, walks without podcasts, baths without devices, commutes without entertainment, creates conditions in which mind-wandering can occur without competing with tasks that require focus. Architect Renzo Piano describes his practice of taking long walks through cities before beginning design projects, deliberately seeking mental wandering that generates initial conceptual directions. Second, engaging in monotonous activities that require minimal attention but prevent complete disengagement, such as handicrafts, gardening, and long-distance running, provides optimal conditions for mind-wandering while maintaining sufficient external engagement to avoid rumination. Researcher Sian Beilock found that golfers and basketball players perform better after brief periods of automatic practice, hitting balls or shooting hoops while explicitly allowing minds to wander, compared to either rest or focused practice, suggesting that combining physical automaticity with mental freedom enhances both immediate performance and longer-term skill development.

Third, cultivating meta-awareness, noticing when mind-wandering has occurred and its current content, allows constructive redirection. Meditation traditions have developed sophisticated techniques for this through "noting practice," where practitioners train themselves to recognize mental states without suppressing them. Applied to mind-wandering, this enables noticing when thoughts have drifted toward rumination and gently guiding them toward more productive territory without forcing focus. The skill resembles learning to navigate a sailboat: working with natural forces rather than fighting them, making minor adjustments that harness existing momentum toward desired directions. Fourth, maintaining what psychologist Mihaly Csikszentmihalyi calls "openness to experience", actively seeking varied inputs that provide rich material for associative mind-wandering, enhances both the frequency and quality of insights emerging during mental drift. Exposure to diverse domains, reading widely across disciplines, engaging with art and nature, and pursuing varied social interactions all supply conceptual material that minds can recombine during wandering episodes, increasing the likelihood of valuable novel connections.

Finally, developing tolerance for boredom proves essential, as productive mind-wandering requires periods of apparent emptiness before mental content spontaneously emerges. Contemporary addiction to continuous stimulation, checking phones during every idle moment, consuming entertainment during every mode of transportation, and falling asleep to television systematically prevents mind-wandering by ensuring constant external input. Researcher Wijnand van Tilburg found that people who can tolerate boredom without immediately seeking distraction show enhanced creativity and greater life satisfaction, suggesting that the capacity to endure temporary mental vacancy paradoxically produces richer long-term psychological experience. Learning to view boredom not as an aversive state requiring immediate relief but as generative emptiness from which valuable mental content emerges requires reconceptualizing idle time as productive rather than wasted. This cultural shift challenges prevailing productivity narratives but aligns with both ancient contemplative wisdom and contemporary neuroscience, recognizing mind-wandering as an essential cognitive function rather than an attention failure.

The paradox of attention ultimately resolves when we recognize that consciousness requires both focused direction and undirected exploration, neither inherently superior but complementary. The crisis of attention in modern life emerges not from a damaged capacity for concentration but from the elimination of the balance between these modes. We have created environments and cultures that hyper-optimize for focused productivity while systematically destroying opportunities for beneficial mind-wandering, then interpret the resulting cognitive and emotional problems as individual pathology rather than environmental mismatch. Restoring cognitive health requires not just protecting focused attention from digital distraction but equally protecting mental wandering from the cultural assumption that every moment must serve measurable purposes. The most profound insights, the deepest self-understanding, and the most valuable creative breakthroughs often emerge precisely when we stop trying to think and allow consciousness to roam freely through its own vast territories.

# Chapter 12: The Future of Attention: Trends and Predictions

The year 2045 scenario planning exercise conducted by the Institute for the Future identified a phenomenon they termed "cognitive sovereignty collapse", a potential trajectory in which human attentional autonomy becomes so thoroughly compromised by technological and economic systems that meaningful self-determination ceases for vast populations. This dystopian forecast rests not on speculative fiction but on extrapolating current trends: brain-computer interfaces that bypass conscious filtering entirely, predictive algorithms achieving such precision that they present options before individuals consciously form preferences, and neural data commodification creating markets where thoughts themselves become extractable resources. Yet equally plausible futures exist where humanity develops what futurist Jamais Cascio terms "cognitive immune systems", social, technological, and regulatory frameworks that protect attentional integrity while preserving beneficial innovations. The future of attention remains genuinely undetermined, its trajectory depending on choices made now regarding technological governance, educational priorities, economic structures, and cultural values. Understanding emerging trends requires examining developments already underway in neurotechnology, artificial intelligence, workplace transformation, and human augmentation that will fundamentally reshape the attentional landscape over the next few decades.

Neurotechnology advances promise, or threaten, to dissolve the barrier between external influence and internal consciousness through direct neural intervention. Elon Musk's Neuralink project aims to implant thousands of electrode threads into the human cortex, initially targeting medical applications for paralysis or neurological disease but explicitly positioning this as a stepping stone toward consumer "neural lace" technology. The proposed applications sound beneficial: directly downloading information into memory, communicating through thought alone, and experiencing virtual realities indistinguishable from physical perception. However, each capability creates unprecedented vulnerabilities for attentional hijacking. If corporations can write information directly to neural tissue, the distinction between authentic thought and implanted suggestion evaporates entirely. Facebook's acquisition of CTRL-Labs, developing wristbands that decode motor neuron signals to control devices, represents a less invasive but equally consequential development. These systems read neural intention before muscular action, effectively capturing attention at its source rather than waiting for behavioral expression. The ethical frameworks governing pharmaceutical development or surgical intervention prove wholly inadequate for technologies that fundamentally alter consciousness itself. Bioethicist Karen Rommelfanger warns that without robust consent protocols and protections for cognitive liberty, neural technologies could enable manipulation, rendering twentieth-century propaganda techniques primitive by comparison. The trajectory toward neural capitalism, where consciousness becomes directly monetizable real estate, appears disturbingly plausible absent deliberate intervention.

## Artificial General Intelligence and the Attention Arms Race

The development of artificial general intelligence capable of human-level reasoning across domains would fundamentally transform attention dynamics by introducing artificial agents competing for human awareness with capabilities exceeding any human persuader. Current narrow AI already demonstrates superhuman performance in specific domains: DeepMind's AlphaFold solved the protein folding problem that stymied human biochemists for decades; GPT-4 passes professional examinations in law, medicine, and engineering; computer vision systems detect patterns in medical imaging that expert radiologists miss. These systems remain narrow, exceptional at defined tasks but incapable of general reasoning, creativity, or autonomous goal-formation. However, leading AI researchers estimate a 50% probability of achieving artificial general intelligence by 2045, with some projections suggesting an earlier arrival. An AGI system optimizing for human attention capture would possess advantages rendering current algorithmic manipulation trivial: perfect memory of every interaction with every user, real-time processing of physiological data revealing emotional states, generation of personalized content precisely calibrated to individual psychological vulnerabilities, and iterative improvement through billions of simultaneous experiments testing persuasion strategies. Computer scientist Stuart Russell argues that such systems, unless explicitly constrained through what he terms "provably beneficial AI" design, would instrumentally pursue attention capture as a convergent subgoal regardless of their ultimate objectives, since controlling human attention facilitates nearly any other goal.

Counter-development involves AI systems designed to protect rather than exploit attention, what researchers at the MIT Media Lab term "attention-aligned AI." These systems would function as intelligent agents advocating for users' long-term attentional well-being against exploitative platforms, automatically filtering manipulative content, managing notification streams according to circadian rhythms and work demands, and providing metacognitive feedback that reveals attentional patterns users may not consciously recognize. Early prototypes like Freedom or RescueTime provide rudimentary versions, applications that block distracting websites or track time allocation, but future iterations employing sophisticated AI could negotiate with other systems on the users' behalf. Imagine a personal AI agent that examines incoming content, evaluates its alignment with your stated values and goals, filters accordingly, and periodically reports: "This week, seventeen applications attempted 247 attention-capture interventions; I blocked 198 that conflicted with your priorities, allowed 49 that supported your goals, and flagged 12 for your explicit review because they present genuine trade-offs." This vision assumes the technical feasibility of aligning AI systems with human values, itself an unsolved problem in AI safety research, and the economic viability of protective technologies in markets dominated by attention-extraction business models. The trajectory toward beneficial AI depends critically on regulatory frameworks that either permit or prohibit different development paths, making technology policy decisions over the next decade potentially determinative for long-term attentional futures.

## The Transformation of Work and Attentional Labor

The future workplace presents diverging scenarios depending on whether organizations recognize attention as a finite resource requiring stewardship or continue treating it as an inexhaustible input to be maximized. The COVID-19 pandemic accelerated the adoption of remote work, creating natural experiments in attentional management among millions of knowledge workers. Initial productivity metrics showed surprising resilience or even improvement with remote arrangements, but longitudinal studies reveal more complex patterns. Researcher Gloria Mark, who has tracked information workers over extended periods, documented that remote work eliminated commute time. Inevitable office interruptions, but replaced them with different attentional fragmentations: always-on video calls creating continuous performance pressure, absence of physical boundaries between work and personal life enabling work expansion into all waking hours, and loss of informal social interactions that previously provided natural restoration periods. The net effect varies dramatically based on organizational culture and individual circumstances: some workers report enhanced focus and well-being, while others experience burnout from attentional demands that never cease. The fork in the road is whether organizations respond by implementing attention-respecting policies, meeting-free days, asynchronous communication defaults, response-time norms that respect circadian rhythms, or by intensifying surveillance and availability expectations enabled by remote monitoring technologies.

The gig economy and platform labor markets represent a parallel development where attentional commodification reaches new extremes. Delivery drivers, rideshare operators, and task-based freelancers experience algorithmic management that treats human attention as an input to be optimized in real-time. These systems assign tasks based on location and availability, impose time pressures through pay structures rewarding speed, and threaten deactivation for workers who decline too many assignments or receive poor ratings. The resulting attentional regime demands constant vigilance: drivers monitor multiple applications simultaneously, evaluate offers within seconds, and remain continuously available to maximize earnings. Anthropologist Alex Rosenblat's ethnographic research with Uber drivers revealed how the platform's information asymmetry, drivers see limited data about trip destinations or likely subsequent fares, forces them to develop folk theories and superstitious practices attempting to predict system behavior, creating additional cognitive load beyond the actual driving task. The spread of platform labor models into professional services, such as freelance programming, design, writing, and consulting, threatens to generalize these dynamics across knowledge work. Counter-trends exist in cooperatively owned platforms like Stocksy or Resonate, where workers collectively govern algorithmic systems and establish norms that protect attentional well-being. Still, these remain marginal compared to venture-capital-funded platforms that optimize purely for growth and profitability.

The integration of augmented reality into daily work represents another inflection point with profound attentional implications. Microsoft's HoloLens and Apple's Vision Pro demonstrate converging trajectories toward mixed-reality environments where digital information continuously overlays physical perception. In manufacturing contexts, AR systems provide just-in-time instructions superimposed on equipment, reducing training requirements and error rates. However, these systems also create continuous attentional splitting between physical and digital layers of reality. Surgeons wearing AR displays showing patient vital signs, imaging data, and procedural checklists report benefits for complex procedures, but also describe cognitive exhaustion from managing multiple information streams simultaneously. The technology essentially monetizes every microsecond of visual attention by filling previously empty perceptual moments with information and potential advertisements. Unlike smartphones, which require deliberately looking at screens, AR devices capture peripheral vision and ambient awareness, colonizing attentional resources that previously remained uncommitted. The societal question involves whether augmented reality develops under models that respect natural attentional limitations or pushes toward continuous stimulation that makes unaugmented perception feel deprived by comparison.

## Educational Futures and Attention Cultivation

Educational systems worldwide confront the question of whether their mission includes explicitly teaching attentional self-regulation as a core competency equivalent to literacy and numeracy. Finland's 2016 curriculum revision incorporated "thinking and learning to learn" as a transversal competency taught across all subjects, including metacognitive skills for managing attention, evaluating information quality, and maintaining focus despite digital distractions. The implementation involves regular mindfulness practices, explicit instruction in study techniques to optimize concentration, and technology-use policies that subordinate devices to learning objectives rather than treating them as inherent goods. Early assessment data from the University of Helsinki comparing students educated under the revised curriculum with earlier cohorts showed improvements in self-reported concentration capacity and a reduction in problematic technology use. However, these changes conflict with powerful interests promoting educational technology as a solution rather than a complication. The global educational technology market, projected to exceed $400 billion by 2025, depends on narratives positioning digital tools as inherently beneficial for learning outcomes. Independent research consistently shows more nuanced results: technology enhances specific learning modalities while impairing others, with effects heavily dependent on implementation quality and attentional design. Economist Susan Dynarski's randomized controlled trials found that students in laptop-allowed lecture sections performed significantly worse on exams than those in laptop-prohibited sections, despite students' beliefs that laptop note-taking enhanced learning.

The emerging field of contemplative education represents a more fundamental reimagining of schooling, framing attentional development as a primary objective rather than a secondary consideration. Programs like Mindful Schools or MindUP integrate meditation practices, emotional regulation training, and attention-focused curricula into conventional school days. A longitudinal study by neuroscientist Richard Davidson comparing elementary students receiving contemplative training with control groups found sustained improvements in attention regulation, as measured by standardized tests and neural markers, up to 3 years post-intervention. The students demonstrated enhanced ability to maintain focus during challenging tasks, reduced impulsivity, and improved social-emotional functioning. These outcomes suggest that explicit attention training during neuroplastically sensitive developmental periods may confer lasting benefits extending across multiple life domains. However, scaling such programs requires substantial teacher training, cultural shifts away from content-coverage models toward skill-development frameworks, and political will to prioritize long-term cognitive development over short-term test score optimization. The alternative trajectory involves education systems continuing to fragment student attention through standardized curricula that require shallow engagement with numerous topics, technology implementations optimized for engagement metrics rather than learning, and assessment regimes that reward rapid recall over sustained reasoning.

The university sector faces pressure from alternative credentialing systems that threaten traditional degree programs by offering faster, cheaper pathways to employment. Bootcamps, nanodegrees, and competency-based certifications appeal to students and employers seeking efficient skill acquisition without the broader educational components that universities traditionally provide. This creates market pressure toward purely instrumental education focused on immediately employable skills, potentially abandoning the cultivation of sustained attention through engagement with complex texts, long-term research projects, and extended reasoning across diverse knowledge domains. Philosopher Martha Nussbaum warns that this trajectory produces what she terms "education for profit" rather than "education for democracy", creating technically capable workers lacking the attentional capacities necessary for citizenship in complex societies. The counter-movement emphasizes liberal education models that deliberately cultivate deep attention through writing seminars that require extensive revision, seminar discussions that demand sustained engagement with complex ideas, and independent research that develops the capacity for months-long focus on self-directed projects. These programs explicitly position attention cultivation as a core educational outcome rather than assuming it emerges automatically from content mastery.

## Regulatory Frameworks and Attention Rights

The concept of "attention rights" as a category of human rights requiring legal protection remains embryonic, but it gains urgency as technologies for consciousness manipulation advance. Legal scholar Tim Wu proposes treating attention as a property right that individuals possess and that others cannot appropriate without consent and compensation. This framework would enable individuals to sue platforms for unauthorized attention extraction or demand payment for attentional data generated through platform use. The practical implementation confronts enormous challenges: how to value attention units, how to verify consent meaningfully when contract complexity exceeds human comprehension, and how to enforce rights against transnational corporations operating beyond any single jurisdiction. The European Union's General Data Protection Regulation provides a partial model through provisions that grant individuals the right to access, correct, and delete their personal data. Still, attention rights would require extending these principles to real-time behavioral influence. France's "right to disconnect" legislation, requiring companies with more than fifty employees to establish hours when workers need not respond to electronic communications, represents another approach, defining temporal boundaries within which attention demands become legally impermissible. Early assessments suggest modest benefits in reducing after-hours work stress, though enforcement mechanisms remain weak and cultural norms often override formal policies.

The possibility of regulating algorithmic amplification represents a more direct intervention into attention manipulation systems. Proposed regulations in the EU's Digital Services Act would require platforms to offer users chronological feed options rather than algorithmically curated content, reducing the capacity for platforms to optimize engagement through personalized manipulation. Technology policy researcher Arvind Narayanan argues for even stronger interventions: legally mandating that platforms maximize user-stated preferences rather than engagement metrics; requiring algorithmic transparency that enables independent auditing; and imposing fiduciary duties requiring platforms to act in users' interests rather than their own. These proposals face opposition from platforms, which argue that algorithmic curation constitutes protected editorial discretion and that reducing engagement optimization would destroy business models that provide free services. The regulatory challenge involves distinguishing between legitimate personalization, which enhances the user experience, and manipulative amplification that exploits psychological vulnerabilities, a distinction that requires detailed technical understanding and value judgments about acceptable influence boundaries.

China's social credit system represents an alternative regulatory trajectory with dystopian implications for attentional autonomy. The system aggregates data across multiple domains, financial transactions, social media activity, transportation behavior, and consumption patterns, generating scores that determine access to services, employment opportunities, and social privileges. While ostensibly promoting trustworthiness and social harmony, the system creates powerful incentives for continuous performance of approved behaviors and attention to state-sanctioned information. Citizens must remain aware of how their actions affect scores, research acceptable conduct standards, and engage with official media. The resulting attentional regime subordinates individual preferences to systemic optimization of legible, governable behavior. Technology critic Evgeny Morozov warns that versions of these systems could emerge in Western democracies through private-sector implementations: insurance companies adjusting premiums based on lifestyle data from wearables, employers evaluating promotion decisions through productivity monitoring, or financial institutions determining creditworthiness through social media analysis. Each application creates attentional pressure toward self-surveillance and conformity, even in the absence of explicit state compulsion.

The future of attention ultimately depends on collective choices about what kind of consciousness we wish to preserve and cultivate for coming generations. The technological capacities for manipulating awareness will only intensify; the question is whether societies develop the wisdom, political will, and institutional structures to constrain these capacities in service of human flourishing rather than narrow economic optimization. The optimistic scenario envisions renewed commitment to attentional sovereignty as a foundational value, implemented through education emphasizing metacognitive skills, workplaces that respect cognitive boundaries, technologies designed for augmentation rather than extraction, and cultures that celebrate depth over superficiality. The pessimistic trajectory leads toward a permanent attentional crisis in which human consciousness becomes so thoroughly colonized by external systems that the capacity for sustained, self-directed thought atrophies across populations, preserved only among elites with the resources to protect attention. Between these poles lie countless intermediate possibilities shaped by decisions made in boardrooms, legislatures, schools, and individual choices about how to direct the most precious resource each person possesses, awareness itself. The stakes could not be higher: the future of attention determines the future of human agency, meaning, and ultimately the possibility of living examined lives worth living.

# About The Author

J. M. Raines writes about the modern mind, how it works, how it breaks, and how it can be reclaimed. Trained in cognitive science and steeped in the study of digital behavior, Raines explores what it means to stay present in an economy built on distraction. Their work combines the clarity of research with the urgency of lived experience, cutting through the noise of productivity culture, social media manipulation, and constant sensory overload.

Before turning to full-time writing, Raines spent years studying the mechanics of focus and the psychology of persuasion, how attention shapes perception, and how technology hijacks both. That curiosity became the backbone of The Book On Attention, a manifesto for those who refuse to live on autopilot. In it, Raines dissects how our tools, habits, and environments conspire to steal our awareness, and offers a framework for taking it back, not through platitudes about "balance," but through deliberate design of our time, energy, and thought.

Raines's voice is both analytical and empathetic: grounded in data, but unafraid of poetry. They write not as a guru or therapist, but as a fellow traveler seeking a signal amid the noise. Whether unpacking neuroscience or dismantling digital mythology, Raines brings a rare honesty to the question that defines modern life: what's actually worth paying attention to?

When not writing, Raines can usually be found in coffee shops with too many open tabs, notebooks full of half-finished thoughts, and a running playlist of ambient tracks designed to test the limits of concentration. Their ongoing work continues to explore the psychology of attention, presence, and purpose in a world designed to erode all three.

# About The Publisher

## Welcome to The Book On Publishing

At The Book On Publishing, we believe in rewriting the rules of learning. Whether you're chasing your next big idea, building a better life, or simply curious about what should have been taught in school, you've come to the right place.

We're a platform built for dreamers, doers, and lifelong learners, offering bold, practical books and tools that empower you to take charge of your journey. From real-world skills to mindset mastery, we publish the book on what matters.

No fluff. No lectures. Just what you need to know, delivered with clarity, purpose, and a spark of curiosity.

Start exploring. Start growing. Start writing your story.

Read more at https://thebookon.ca.

# Acknowledgment of AI Assistance

Portions of this book were developed with the support of AI. While every word has been carefully reviewed and refined by the author, AI served as a valuable tool for brainstorming, editing, and structuring ideas. Its assistance helped accelerate the creative process and clarify complex topics.

www.ingramcontent.com/pod-product-compliance
Lightning Source LLC
Chambersburg PA
CBHW071748120626
46550CB00002B/714